A-Level Mathi ||||| Year 1 Es _____ntials

The Colour Guide to What You Need to Know
Christine Norledge

This book is a stunning and mathematically accurate summary of what you need to know for Year 1 A-Level Mathematics. Chapters are linked to the topics of Tarquin A Level textbook – designed to be used with any Board. As a summary of each topic, no better exists – perfect for revision, for checking understanding or even putting on a wall to remind you of key facts. For linked resources, see the back cover. To add your own notes and print as a notes or as a poster, see the Tarquin eReader version.

Contents

Published by Tarquin Publications
Suite 74, 17 Holywell Hill
St Albans AL1 1DT, UK

www. tarquingroup.com
Printed and designed in the United Kingdom

Copyright © Christine Norledge, 2018
Book ISBN: 978-1-91109376-3
Ebook ISBN 978-1-91109377-0

Indices & Surds

Laws for integer indices:

$$b^m \times b^n = b^{m+n}$$
$$b^m \div b^n = b^{m-n}$$
$$(b^m)^n = b^{mn}$$
$$b^0 = 1 \quad (b \neq 0)$$

Reciprocal →
$$b^{-m} = \frac{1}{b^m} \quad (b \neq 0)$$

A negative power indicates a reciprocal:
$$4^{-3} = \frac{1}{4^3} = \frac{1}{64}$$

$$\left(\frac{3}{4}\right)^{-2} = \left(\frac{4}{3}\right)^2 = \frac{16}{9}$$

Fractional indices represent roots:
$$x^{\frac{1}{2}} = \sqrt{x}$$
$$x^{\frac{1}{q}} = \sqrt[q]{x}$$
$$x^{\frac{p}{q}} = \sqrt[q]{x^p}$$

A negative fractional power is treated as follows

$$16^{-\frac{3}{2}} = \left(\frac{1}{16}\right)^{\frac{3}{2}} = \sqrt{\left(\frac{1}{16}\right)^3} = \left(\frac{1}{4}\right)^3 = \frac{1}{64}$$

Deal with reciprocal

Deal with root first

Applications

Laws of indices are useful for simplifying expressions.

Surds are used when an exact answer is needed, particularly when working with quadratics or trigonometry.

Surds

A surd denotes an irrational root

i.e. $\sqrt{9} = 3$ (rational)
$\sqrt{5} \approx 2.236$ (irrational)

Laws of surds:
$$\sqrt{x} \times \sqrt{y} = \sqrt{xy}$$
$$\sqrt{x} \div \sqrt{y} = \sqrt{\frac{x}{y}}$$

To simplify a surd, extract the largest square factor:
$$\sqrt{405} = \sqrt{81 \times 5}$$
$$= \sqrt{81} \times \sqrt{5}$$
$$= 9\sqrt{5}$$

Power equations

To solve equations using powers, ensure tha each side has the same base:

$$3^{3x} = 9^{x+2}$$
$$= (3^2)^{x+2}$$
$$= 3^{2x+4}$$
$$\Rightarrow 3x = 2x + 4$$
$$\Rightarrow x = 4$$

Here, rewrite 9 as 3 squared an use base of 3 throughout

Simplifying surds

To add or subtract surds, rewrite using a common surd:

$$\sqrt{45} + \sqrt{20} = \sqrt{9} \times \sqrt{5} + \sqrt{4} \times \sqrt{5}$$
$$= 3\sqrt{5} + 2\sqrt{5}$$
$$= 5\sqrt{5}$$

Expressions with brackets can be expanded:

$$(3\sqrt{3} + 2)(4 - 2\sqrt{3}) = 12\sqrt{3} + 8 - 6 \times 3 - 4\sqrt{3}$$
$$= 8\sqrt{3} - 10$$

Rationalising the denominator

It is better to write surds involving fractions with a rational (non-surd) denominator.

In simple cases, multiply numerator and denominator by a suitable surd:

$$\frac{2\sqrt{3}}{\sqrt{7}} = \frac{2\sqrt{3}\sqrt{7}}{7} = \frac{2\sqrt{21}}{7}$$

$$\sqrt{7} \times \sqrt{7} = 7$$

In more complicated cases, we apply the difference of two squares to ensure that th denominator will simplify to a rational number:

Difference of two squares

$$\frac{3\sqrt{5}}{4 + 2\sqrt{5}} = \frac{3\sqrt{5}\,(4 - 2\sqrt{5})}{(4 + 2\sqrt{5})(4 - 2\sqrt{5})}$$
$$= \frac{12\sqrt{5} - 6 \times 5}{16 - 4 \times 5}$$
$$= \frac{12\sqrt{5} - 30}{-4} = \frac{15}{2} - 3\sqrt{5}$$

Vectors

Applications

- In geometry: finding points of intersection and distances between points.
- In mechanics: motion of particles and resolving forces.
- Vectors are also useful for modelling systems in 3D.

What is a vector?

▷ Vectors and scalars

- A vector represents motion in a particular direction (e.g. force, velocity).
- A scalar simply has a magnitude (size), and the direction is not specified (e.g. distance, speed).

▷ Notation and diagrams

- This represents the vector $a = \overrightarrow{AB}$:

Line length = magnitude
Direction shown by arrow

- Vectors are typed in **bold**, but underlined when handwritten.

Magnitude

- The magnitude of **a** is written $|a|$ and can be found using Pythagoras' theorem.

$$a = \begin{pmatrix} 3 \\ 4 \end{pmatrix}$$
$$\Rightarrow |a| = \sqrt{3^2 + 4^2}$$
$$= \sqrt{25} = 5$$

Unit vector

- A unit vector $\hat{a}$ in the direction of vector **a** is:

$$\hat{a} = \frac{a}{|a|}$$

- e.g:

$$a = \begin{pmatrix} 3 \\ 4 \end{pmatrix}$$
$$\Rightarrow \hat{a} = \frac{a}{|a|} = \frac{3i+4j}{5} = \frac{3}{5}i + \frac{4}{5}j$$

- A unit vector has magnitude = 1.

Vector basics

▷ Reverse vectors:

- If $a = \overrightarrow{AB}$, then $-a = \overrightarrow{BA}$.

▷ Equality and parallel vectors:

- Two vectors are equal ($a = b$) if they have the same magnitude and direction.
- Two vectors are parallel if one is a scalar multiple of the other. (e.g. $a = 3i + 4j$ and $b = 6i + 8j$ are parallel because $b = 2a$)

▷ The zero vector:

- **0** has no magnitude and arbitrary direction.

Vector components

- In 2D, the vectors **i** and **j** represent unit vectors in the directions x and y respectively.

  magnitude = 1

- For the vector $a = 3i + 4j$, the scalars 3 and 4 are the components of the vector **a**.

  ▷ Row form: $a = (3,4)$ ▷ Column form: $a = \begin{pmatrix} 3 \\ 4 \end{pmatrix}$

- Working from the origin 0, the point with coordinates $(3,4)$ has **position vector** $r = 3i + 4j$.

  r frequently used to denote position vector

Vector calculations

- To add or subtract vectors, add or subtract their components:

$$a = \begin{pmatrix} 5 \\ 3 \end{pmatrix}, b = \begin{pmatrix} 3 \\ -2 \end{pmatrix}$$
$$a + b = \begin{pmatrix} 8 \\ 1 \end{pmatrix}$$

- Vector addition is associative:

$$a + (b + c) = (a + b) + c$$

- Scalar multiplication is commutative:

$$k(b + c) = kb + kc$$

Linear Functions

Applications

- Modelling linear relationships (e.g. profit and loss to find break-even point).
- Solving problems using coordinate geometry.
- Skills developed in this chapter are fundamental to A-level Maths.

Cartesian coordinates

- The 2D Cartesian coordinate system describes each point with an ordered pair (x,y).
  - x-coordinate = abscissa
  - y-coordinate = ordinate
- The Cartesian plane is divided into four quadrants.

- We can define two general points on a line: A (x_a, y_a) and B (x_b, y_b).

Gradient

- The gradient of a line joining two points A and B is

$$m = \frac{y_b - y_a}{x_b - x_a}$$

- For the points $(3,-6)$ and $(5,2)$:

$$m = \frac{2--6}{5-3} = \frac{8}{2} =$$

Parallel & perpendicular

- Two lines l_1 and l_2 are parallel if they have the same gradient.

- Two lines l_1 and l_2 are perpendicular if the product of their gradients is -1:

$$m_1 m_2 = -1 \Rightarrow m_2 = \frac{-1}{m_1}$$

Finding the equation of a line

- Using the gradient and one point on the line:

$$y - y_a = m(x - x_a)$$

- For m = -2 and the point (4,6):

$$y - 6 = -2(x - 4)$$
$$\Rightarrow y = -2x + 14$$

- Using two points on the line:

$$\frac{y - y_a}{y_b - y_a} = \frac{x - x_a}{x_b - x_a}$$

- For the points (3,-6) and (5,2):

$$\frac{y - -6}{2 - -6} = \frac{x - 3}{5 - 3}$$

$$\Rightarrow \frac{y + 6}{8} = \frac{x - 3}{2}$$

$$\Rightarrow 2(y + 6) = 8(x - 3)$$

$$\Rightarrow y = 4x - 18$$

Forms of a linear function

- The equation of the line with gradient m and y-intercept c can be written in the form:

$$y = mx + c$$

- It can also be written in the form:

$$ax + by + d = 0$$

  - This second form can be useful when m or c are fractions.
  - Substitute x = 0 or y = 0 to find the intercepts:

$$x = 0 \Rightarrow y = \frac{-d}{b}$$
$$y = 0 \Rightarrow x = \frac{-d}{a}$$

  - The gradient is given by:

$$m = \frac{-a}{b}$$

Distance between points

- Use Pythagoras' theorem to find the distance between two points.

- For the distance between points A and B:

$$|AB| = \sqrt{(x_b - x_a)^2 + (y_b - y_a)^2}$$

- For the points A = (3,-6) and B = (5,2):

$$|AB| = \sqrt{(2 - 5)^2 + (-6 - 3)^2}$$
$$= \sqrt{90} = 3\sqrt{10}$$

Midpoint of a line segment

- Find the "average" of the two points.

- For the points A and B, the coordinates of the midpoint are given by:

$$\left(\frac{x_a + x_b}{2}, \frac{y_a + y_b}{2} \right)$$

- For the points A = (3,-6) and B = (5,2):

$$\left(\frac{3+5}{2}, \frac{-6+2}{2} \right) = (4, -2)$$

Quadratic Functions

General quadratic

$$f(x) = ax^2 + bx + c$$

- a, b and c are real.
- a is not 0.
- When $a = 1$, the quadratic is called monic.

Applications

▷ Fundamental to many parts of A-level Maths.

▷ Quadratics appear when modelling many physical situations mathematically.

▷ i.e. projectile moving through air (parabola) or problems involving area.

Graph shape

$a > 0$

$a < 0$

Factorise and solve

- ▨ Monic quadratics

 ● Find two factors of c that sum to give b.

$$x^2 + 3x - 10$$
$$\Rightarrow (x - 2)(x + 5)$$
$$\Rightarrow x = 2 \text{ and } x = -5$$

- ▨ General quadratics

 ● Use inspection.
 ● Consider factor pairs.

$$3x^2 - 10x - 8 = 0$$
$$\Rightarrow (3x + 2)(x - 4) = 0$$
$$\Rightarrow x = -\frac{2}{3} \text{ and } x = 4$$

- ▨ Most quadratics won't factorise - but try first!

The discriminant

$$\Delta = b^2 - 4ac$$

Gives us information about the roots of a quadratic.

▷ $\Delta > 0$ Two real distinct roots; graph intersects x axis twice.

▷ $\Delta = 0$ One repeated root; graph intersects x axis once.

▷ $\Delta < 0$ No real roots; graph does not intersect x axis.

Complete the square

- ▨ Monic quadratics

 ● Complete the square then rearrange to solve.

$$x^2 - 4x - 5 = 0$$
$$\Rightarrow (x - 2)^2 - 9 = 0$$
$$\Rightarrow x = 9 \pm \sqrt{2}$$

- ▨ General quadratics
 ● Remove factor of a and proceed as for monics.
- ▨ Useful for:
 ● Exact solution when quadratic won't factorise.
 ● Finding turning point for sketching graph.

$$(x - 2)^2 - 9$$
$$\Rightarrow \text{turning point at } (2, -9)$$

Sketching the graph

● The roots of a quadratic are the solutions to
↓
The x-intercept(s)

$$ax^2 + bx + c = 0$$

● To find the coordinates of the turning point, complete the square.
↓
Maximum or minimum

● Find the y-intercept by substituting $x = 0$ and label.

y-intercept

roots

turning point

Quadratic formula

- ▨ Derived by completing the square on the general quadratic.

- ▨
$$x = \frac{-b \pm \sqrt{b^2 - 4ac}}{2a}$$

- ▨ Can be used to solve any quadratic (but be careful as some quadratics do not have real roots).

Polynomials

A polynomial is an expression of the form: $a_0 + a_1x + a_2x^2 + \ldots + a_nx^n$

- n is an integer (the degree of the polynomial).
- $a_0, a_1, \ldots, a_n$ are constants, $a_n \neq 0$.
- Quadratics are polynomials of order 2 and cubics are polynomials of order 3.

Graphs of polynomials

▷ Use the degree of the polynomial (n) to help to sketch the graph:

☐ If n is even:

$x^n \longrightarrow \infty$ as $x \longrightarrow -\infty$

$x^n \longrightarrow \infty$ as $x \longrightarrow \infty$

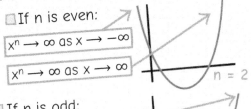

$n = 2$

☐ If n is odd:

$x^n \longrightarrow \infty$ as $x \longrightarrow \infty$

$x^n \longrightarrow -\infty$ as $x \longrightarrow -\infty$

$n = 3$

☐ A polynomial of degree n will have no more than $n - 1$ turning points (e.g. a quadratic ($n = 2$) has one turning point)

▷ If the coefficient $a_n < 0$, graph shape is reflected in the x-axis.

Applications

▷ Manipulation of polynomials is important for further aspects of the A-level Maths course.

▷ Polynomials are used to approximate complicated curves in CAD or computer graphics.

Roots

◉ The root of a polynomial is where it crosses the x-axis: $f(x) = 0$

◉ A polynomial of degree n has at most n different real roots.

Multiplying polynomials

▨ Multiply every term in the first polynomial by every term in the second polynomial:

$(2x^2 + 3x - 5)(4x + 2)$
$= 8x^3 + 12x^2 - 20x + 4x^2 + 6x - 10$
$= 8x^3 + 16x^2 - 14x - 10$

Factor theorem

◉ The expression $(x - b)$ is a factor of the polynomial $f(x)$ if and only if $f(b) = 0$.

◉ In other words, if we can find a root b, we know that $(x - b)$ must be a factor.

◉ e.g. $f(x) = 2x^3 + x^2 - 15x - 18$
$f(3) = 2(3)^3 + (3)^2 - 15\times3 - 18 = 0$

$\Rightarrow x = 3$ is a root
$\Rightarrow (x - 3)$ is a factor of $f(x)$

Dividing polynomials

▨ Dividing polynomials is useful for finding factors, and therefore roots, of a polynomial.

▨ Factor theorem can be used to find one factor; use long division/equating coefficients to find other factors.

$(x - 3)(ax^2 + bx + c) = 2x^3 + x^2 - 15x - 18$

Coefficient of $x^3 \Rightarrow a = 2$
Constant $\Rightarrow -18 = -3c \Rightarrow c = 6$
Coefficient of $x^2 \Rightarrow -3a + b = 1 \Rightarrow b = 7$

$2x^3 + x^2 - 15x - 18 = (x - 3)(2x^2 + 7x + 6)$
$= (x - 3)(2x + 3)(x + 2)$

Remainder theorem

◉ When a polynomial $f(x)$ is divided by $(x - b)$, the remainder is $f(b)$.

Direct Proof

Applications
▷ It is important to be able to show that a mathematical result is always true; this is done using reasoned, mathematical argument (proof).
▷ The concept of proof is fundamental to maths.

Methods of proof

▨ There are many different methods of proof; only some are required at A-level.

▨ Graphical/geometric:
▶ Involves use of a diagram.
▶ e.g. Many proofs of Pythagoras' theorem are graphical.

▨ Deduction:
▶ Follow a series of logical deductions to conclude that a statement must be valid.
▶ e.g. Many proofs encountered at GCSE, such as "prove that the sum of two consecutive numbers is an odd number".

Counter-example

◉ To prove that a statement is incorrect, it is sufficient to find only one counter-example.

◉ e.g. The prime 2 is a counter-example to "all primes are odd".

Methods of proof

▨ Contradiction:
▶ Assume a statement to be false, then show that this leads to a contradiction.
▶ e.g. The proof that the square root of 2 is irrational.

▨ Exhaustion:
▶ Show that a statement is true for all of a finite amount of cases.
▶ e.g. The proof that no square number ends in 8.
▶ This is only practical when there is a relatively small number of cases to consider!

Notation

◉ "A implies B" is written:
$$A \Rightarrow B$$

◉ This can also be written as "A only if B" or "if A, then B".

◉ The following statements are equivalent:
$$B \Rightarrow A$$
$$A \Leftarrow B$$

◉ Two way implication:
If $A \Rightarrow B$ and $B \Rightarrow A$ then $A \Leftrightarrow B$

◉ This can also be written as "A if and only if B" or "A iff B".

Positive integer proofs

▷ The following are useful when proving statements about the positive integers:

☐ The positive integer m is even if there exists a positive integer p such that:
$$m = 2p$$

☐ The positive integer n is odd if there exists a positive integer q such that:
$$n = 2q - 1$$

And/Or statements

◉ "If A and B then C" means that C holds if both A and B hold.

◉ "If A or B then C" means that C holds if at least one of A or B holds.

Binomial Expansion

The expansions of $(x + y)^n$ for n = 2 and 3 are:

$$(x + y)^2 = x^2 + 2xy + y^2$$

$$(x + y)^3 = x^3 + 3x^2y + 3xy^2 + y^3$$

The coefficients of the expansions appear in Pascal's Triangle:

```
        1          ← n = 0
      1   1
    1   2   1
  1   3   3   1
1   4   6   4   1  ← n = 4
↑               ↑
r = 0         r = 4
```

Not efficient for large values of n.

Definition

A binomial is a polynomial that is the sum of two terms:

$$ax^p + bx^q$$

a and b are real

p and q are integers.

Applications

Historically used to calculate high powers of numbers close to 1.

Now particularly useful in Statistics, where it leads to an important statistical distribution - the binomial distribution.

Binomial coefficient

The factorial of n is defined as

$$n! = n(n - 1)(n - 2)...(2)(1)$$

e.g. 3! = 3 x 2 x 1 = 6   0! = 1

n! = n(n - 1)! = n(n - 1)(n - 2)... (2)(1)

The rth entry (counting from r = 0) of the nth row can be calculated using:

$$\binom{n}{r} = \frac{n!}{r!(n - r)!}, \quad r \le n$$

e.g. To find 3rd entry in the 4th row:

NB: Count from r = 0

$$\binom{4}{3} = \frac{4!}{3! \, 1!} = \frac{24}{6} = 4$$

Further points

Alternative notation:

$$\binom{n}{r} = n \, C \, r = {}^nC_r$$

This is said "n choose r" and is used further for the binomial distribution.

This is also the notation used on a calculator.

Symmetry: $\binom{n}{r} = \binom{n}{n - r}$

Each row in Pascal's Triangle is symmetrical.

Pascal's Identity:

$$\binom{n}{r - 1} + \binom{n}{r} = \binom{n + 1}{r}$$

Factorial calculations

Factorial calculations should be simplified if you do not have access to a calculator:

$$\binom{10}{7} = \frac{10!}{3! \, 7!}$$

$$= \frac{10 \times 9 \times 8 \times 7!}{3 \times 2 \times 1 \times 7!}$$

$$= \frac{720}{6} = 120$$

The binomial expansion

The binomial expansion of $(a + bx)^n$ is:

$$(a + bx)^n = \binom{n}{0}a^n + \binom{n}{1}a^{n-1}bx + \binom{n}{2}a^{n-2}(bx)^2 + ... + \binom{n}{n}(bx)^n$$

Don't forget to raise b to the relevant power too!

e.g. The expansion of $(3 + 2x)^4$ is:

$$(3 + 2x)^4 = \binom{4}{0}3^4 + \binom{4}{1}3^3(2x) + \binom{4}{2}3^2(2x)^2 + \binom{4}{3}3^1(2x)^3 + \binom{4}{4}(2x)^4$$

$$= 1(81) + 4(54x) + 6(36x^2) + 4(24x^3) + 1(16x^4)$$

$$= 81 + 216x + 216x^2 + 96x^3 + 16x^4$$

Sometimes the full expansion isn't required. e.g. The coefficient of x^5 in the expansion of $(1 - 2x)^{10}$ is:

$$\binom{10}{5}1^5(-2x)^5 = 252(-32x^5) = -8064x^5$$

We frequently deal with cases where x is sufficiently small that we can disregard higher powers. In exams, you will often be asked to find up to a given power of x (usually x^3 or x^4).

e.g. Expand $(1 + 3x)^6$ up to the term in x^3 and hence find an approximation for 1.003^6.

$$(1 + 3x)^6 \approx \binom{6}{0}1^6 + \binom{6}{1}1^5(3x) + \binom{6}{2}1^4(3x)^2 + \binom{6}{3}1^3(3x)^3$$

$$\approx 1(1) + 6(3x) + 15(9x^2) + 20(27x^3)$$

$$\approx 1 + 18x + 135x^2 + 540x^3$$

Substitute x = 0.001 into our approximation

$$(1 + 0.003)^6 = (1 + 3 \times 0.001)^6$$

$$\approx 1 + 18 \times 0.001 + 135 \times 0.001^2 + 540 \times 0.001^3$$

$$\approx 1.01813554$$

Sketching Graphs

Shapes of polynomials

▷ For a polynomial of degree n:

n = 2
Quadratic
0, 1 or 2
real roots

n = 3
Cubic
1, 2 or 3
real roots

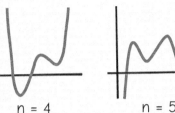

n = 4
Quartic
0, 1, 2, 3 or 4
real roots

n = 5
Quintic
1, 2, 3, 4 or 5
real roots

▷ Odd degree:
  ☐ Polynomial has at least one real root.
  ☐ If number of roots is even, there will be repeated roots.
▷ Even degree:
  ☐ Polynomial may not have real roots.
  ☐ If number of roots is odd, there will be repeated roots.

Applications

◉ Sketching a graph of an equation can be useful to look at the long-term behaviour of a function.

◉ Graph sketches can also be used to check answers or solutions to equations.

Fractional powers

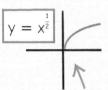

$y = x^{\frac{1}{2}}$

Positive root only - 4th quadrant empty. Happens when denominator of power is even.

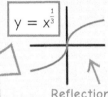

$y = x^{\frac{1}{3}}$

Reflection of $y = x^3$ in line $y = x$

Negative powers

▨ These functions are discontinuous as they are undefined at x = 0 and y = 0. The x- and y-axes are asymptotes.

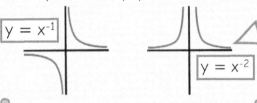

$y = x^{-1}$

$y = x^{-2}$

Direct proportion

▨ If a quantity y is directly proportional to a quantity x, we write: $\boxed{y \propto x}$

▨ This means that there is a (non-zero) constant, k, such that: $\boxed{y = kx}$

▨ The units of the constant, k, should always be stated.

▨ A direct proportion graph passes through the origin.

▨ e.g. The mass (m) of a gold bar is directly proportional to its volume (v).
When m = 500g, v = 25.9cm³ (3sf). Find k:

$$m \propto v$$
$$\Rightarrow m = kv$$
$$\Rightarrow k = \frac{m}{v}$$
$$\Rightarrow k = \frac{500}{25.9} = 19.3 \text{g/cm}^3$$

g/cm³ = density
Hence we have found the density of gold (to 3sf).

Inverse proportion

▨ If a quantity y is inversely proportional to a quantity x, we write: $\boxed{y \propto \frac{1}{x}}$

▨ This means that there is a (non-zero) constant, k, such that: $\boxed{y = \frac{k}{x}}$

Sketching graphs

◉ For polynomials, find roots - this may require application of techniques for solving quadratics or the factor theorem.

◉ Find and label:
  ▷ Points of intersection with the axes by substituting x = 0 and y = 0.
  ▷ The equations of any asymptotes.
  ▷ Turning points if required - remember that you can use completing the square for quadratics.

◉ Points of intersection:
  If more than one graph is sketched on the same set of axes, lines may intersect.
  ▷ Sketching a graph may provide a useful insight into the region(s) of points of intersection; these can then be confirmed using analytic methods.

Transformation of Graphs

Applications

- Graphs arising from real-life scenarios are unlikely to be simple functions.
- We can apply a series of transformations to fit a "standard" curve to a more complex curve.
- e.g. Using a combination of transformations on the standard normal distribution curve to fit real-life data.

Translations

▷ $y = f(x) + a$ translates graph of $y = f(x)$ a units in the y-direction.

e.g. Transformation $y = f(x) + 2$:

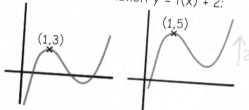

This transformation could be represented using the vector $\begin{pmatrix} 0 \\ 2 \end{pmatrix}$.

▷ $y = f(x + a)$ translates graph of $y = f(x)$ $-a$ units in the x-direction.

e.g. Transformation $y = f(x + 3)$:

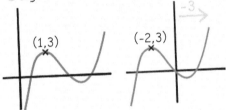

This transformation could be represented using the vector $\begin{pmatrix} -3 \\ 0 \end{pmatrix}$.

▷ Combining these two results in the translation $f(x + 3) + 2$, moving the point (1,3) to (-2,5).

Reflections

Multiplying by -1 causes a reflection in either the x- or y-axis.

$y = -f(x)$ reflects graph in the x-axis.

$y = f(-x)$ reflects graph in the y-axis.

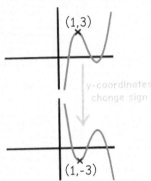
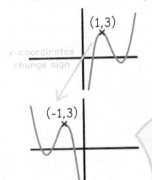

Stretches

$y = af(x)$ stretches the graph of $y = f(x)$ by a factor a parallel to the y-axis.

▶ e.g. Transformation $y = 2f(x)$:

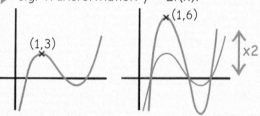

▶ y-coordinates are multiplied by 2.

▶ Notice that the roots do not change position.

Stretches

$y = f(ax)$ stretches the graph of $y = f(x)$ by a factor $1/a$ parallel to the x-axis.

▶ e.g. Transformation $y = f(2x)$:

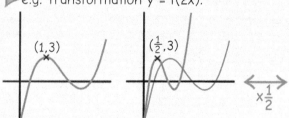

▶ x-coordinates are divided by 2.

▶ Here, the roots change position but the y-intercept remains in the same position.

Summary

- Position of constant:
  - Constant outside brackets affects y-coordinates.
  - Constant inside brackets affects x-coordinates.
- Effects of constant:
  - y-coordinate does what you'd expect (i.e. moves up/multiplies).
  - x-coordinate does the opposite (i.e. moves left/multiplies by reciprocal)
  - Negative constant results in a reflection.
- Label any key points (e.g. turning points).
- Transform any asymptotes.

Simultaneous Equations

Applications

A system of simultaneous equations contains two or more variables.

Solving a system of simultaneous equations allows us to find the intersection point(s) of lines and curves.

These techniques are frequently applied to a variety of problems in pure and applied mathematics.

Elimination

Use operations to produce a single equation in one variable:

$2x + 4y = 12$ ① ← Number each equation
$3x - 2y = 14$ ②

① × 3 ⇒ $6x + 12y = 36$ ③
② × 2 ⇒ $6x - 4y = 28$ ④

Subtracting ④ from ③:
$16y = 8 ⇒ y = \frac{1}{2}$

Substitute $y = \frac{1}{2}$ into ②: ← Either equation can be used
$3x - 2y = 14$
⇒ $3x - 1 = 14$
⇒ $x = 5$

Solution: $(x,y) = (5, \frac{1}{2})$

Solution ⇄ methods

Substitution

Substitute an expression from one equation into another:

$2xy = 16$ ①
$2x - 3y = -8$ ②

② ⇒ $2x = 3y - 8$ ← Look for straightforward rearrangements

Substituting into ①:
$2xy = 16$
⇒ $(3y - 8)y = 16$
⇒ $3y^2 - 8y = 16$
⇒ $(3y + 4)(y - 4) = 0$ ← May need to apply methods for solving quadratics
⇒ $y = -\frac{4}{3}$ or $y = 4$

Substitute each of these into ② to get solutions:
$(x,y) = (-6, -\frac{4}{3})$ and
$(x,y) = (2,4)$

Pairs of linear equations

A pair of simultaneous linear equations will have 0 or 1 intersection points (or infinite intersection points if lines coincide).

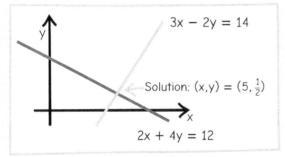

Most pairs of simultaneous linear equations have one solution and intersection point.

A pair of parallel lines never meet, so have no intersection points.

Graphical solutions

Sketching a graph of the system of equations is a good way to check answers.

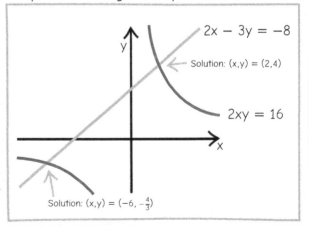

Linear and quadratic

Two POI   One POI   No POI

Linear and circle

Two POI   One POI   No POI

Inequalities

Linear inequalities are solved in the same manner as linear equations and solution sets can be shown on a number line:

$$3x + 5 \geq 11$$
$$\Rightarrow 3x \geq 6$$
$$\Rightarrow x \geq 2$$

▷ A solid circle shows that the value 2 is included.

▷ The solution set is written:
$$\{x: x \geq 2\}$$

Applications

- Simple problems such as working with discriminants.
- Linear programming (i.e. finding the maximum profit to be made from sales).
- Sudoku can be formulated as a linear programming problem.

Double inequalities

$$6 < 2x - 4 \leq 10$$
$$\Rightarrow 10 < 2x \leq 14$$
$$\Rightarrow 5 < x \leq 7$$

▷ An unfilled circle shows that the value 5 is not included.

▷ The solution set is written:
$$\{x: 5 < x \leq 7\}$$

Negatives

When multiplying or dividing by a negative number, reverse the inequality symbol:

$$2(8 - 3x) < 28$$
$$\Rightarrow 16 - 6x < 28$$
$$\Rightarrow -6x < 12$$
$$\Rightarrow x > -2$$

Quadratic inequalities

Quadratic inequalities are also solved in the same manner as quadratic equations:

$$2x^2 - 5x - 12 \leq 0$$
$$\Rightarrow (2x + 3)(x - 4) \leq 0$$
$$\Rightarrow \text{Roots } x = -\frac{2}{3} \text{ and } x = 4$$

Use a graph to find the correct solution set:

Above x-axis $2x^2 - 5x - 12 > 0$

Need this part of the graph.

$(-\frac{2}{3}, 0)$ $(4, 0)$ Below x-axis $2x^2 - 5x - 12 < 0$

Solution: $\{x: -\frac{2}{3} \leq x \leq 4\}$

Simultaneous inequalities

For single-variable simultaneous inequalities, we need to look at which region(s) satisfy both inequalities:

① $3x + 5 \geq 11$
$\Rightarrow \{x: x \geq 2\}$
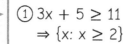

② $2x^2 - 5x - 12 \leq 0$
$\Rightarrow \{x: -\frac{2}{3} \leq x \leq 4\}$
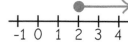

Combining these:
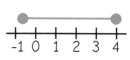

Solution set: $\{x: 2 \leq x \leq 4\}$

▷ e.g. Find the region satisfying: $y \geq 2x + 5$, $y > 2x^2 - 5x - 12$

○ We need the region above the line as before.

○ Checking the point (1,2) inside the parabola: $2 > 2(1)^2 - 5(1) - 12$
$\Rightarrow 2 > -15$
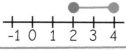

○ This is correct so we need the region inside the parabola and above the line.

Multi-variable inequalities

▷ Inequalities in two variables can be represented using region(s) on a graph.

▷ e.g. Region $y > 2x + 5$:
○ Draw graph of equation $y = 2x + 5$.
○ Test points on either side of the line to work out which side satisfies the inequality.

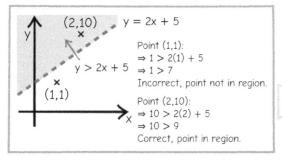

(2,10) $y = 2x + 5$

$y > 2x + 5$

(1,1)

Point (1,1):
$\Rightarrow 1 > 2(1) + 5$
$\Rightarrow 1 > 7$
Incorrect, point not in region.

Point (2,10):
$\Rightarrow 10 > 2(2) + 5$
$\Rightarrow 10 > 9$
Correct, point in region.

▷ Dashed or solid line?
○ A dashed line is used with a strict inequality to show that the points on the boundary are not included.
○ A solid line is used when the points on the boundary are included.

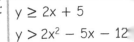
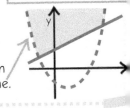
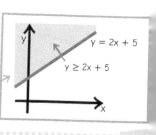

$y = 2x + 5$
$y \geq 2x + 5$

Circle Geometry

Applications

▷ Modelling problems involving geometry of circles (e.g. applications in engineering).

▷ Links to trigonometry through π and relationship between circumference and area.

▷ Links to prior work on linear and polynomial functions (e.g. finding intersection points between circles & lines/curves).

Circle theorems

▨ The angle in a semicircle = 90°.

▷ i.e. If AB is the diameter of a circle and C lies on the circumference, then angle ACB is a right angle.

▨ The angle at the centre of a circle is twice the angle at the circumference.

▷ i.e. If angle ABC = $y°$ then angle AOC = $2y°$.

▨ The perpendicular bisector of a chord of a circle passes through the centre of the circle.

▷ i.e. If AB is a chord of the circle, OC bisects AB at right angles.

▨ A tangent to a circle is perpendicular to the radius.

i.e. If AB is a tangent to the circle (touches at only one point, C), then ACO and BCO are both 90°.

Equation of a circle

◉ The equation of a circle with centre (a, b) and radius r is:

$$(x - a)^2 + (y - b)^2 = r^2$$

◉ e.g. The equation of a circle with centre $(3, 5)$ and radius 4 is:

$$(x - 3)^2 + (y - 5)^2 = 16$$

$4^2 = 16$

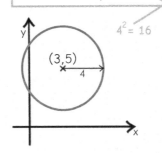

(3,5)

Using Pythagoras

▷ The points A(3,4), B(2,-1) and C(2,5) lie on a circle. Use Pythagoras' theorem to show that the line AB is the diameter of the circle.

A = (3,4), B = (-1,2), C = (2,5)

$$|AB| = \sqrt{(3 - -1)^2 + (4 - 2)^2}$$
$$= \sqrt{20} = 2\sqrt{5}$$

$$|AC| = \sqrt{(3 - 2)^2 + (4 - 5)^2}$$
$$= \sqrt{2}$$

$$|BC| = \sqrt{(-1 - 2)^2 + (2 - 5)^2}$$
$$= \sqrt{18}$$

$$|AC|^2 + |BC|^2 = \sqrt{2}^2 + \sqrt{18}^2$$
$$= 2 + 18 = 20$$
$$= |AB|^2$$

▷ Therefore ABC is right-angled, angle ACB = 90° and so AB is the diameter of the circle.

To find the centre/radius of a circle

◉ To find the centre/radius of a circle when the equation is expanded, complete the square.

◉ e.g. Find the centre & radius:

$$x^2 - 4x + y^2 + 6y = 12$$

$x^2 - 4x + y^2 + 6y = 12$
$\Rightarrow (x - 2)^2 - 2^2 + (y + 3)^2 - 3^2 = 12$
$\Rightarrow (x - 2)^2 + (y + 3)^2 - 13 = 12$
$\Rightarrow (x - 2)^2 + (y + 3)^2 = 25$
$\Rightarrow$ Circle radius 5, centre $(2, -3)$

◉ e.g. The line joining points A(3,4) and B(-1,2) is the diameter of a circle. Find the equation of the circle.

☐ Midpoint AB = centre of circle:

Midpoint of AB = $\left(\frac{3-1}{2}, \frac{4+2}{2}\right) = (1,3)$

☐ Distance AB = length of diameter:

$|AB| = \sqrt{(3 - -1)^2 + (4 - 2)^2}$
$= \sqrt{20} = 2\sqrt{5}$

☐ Combine to form equation:

$\Rightarrow$ Centre = $(1,3)$, radius = $\sqrt{5}$
$\Rightarrow$ Equation is $(x - 1)^2 + (y - 3)^2 = 5$

Proofs

▨ It is important that you understand proofs of the circle theorems.

▨ They are mostly proved geometrically; use diagrams to carefully illustrate steps taken.

▨ The tangent/radius theorem can be proved using calculus.

Applications of linear geometry

▷ Use knowledge of perpendicular gradients to find equations of tangents and bisectors.

▷ e.g. Find the equation of the tangent to circle $(x - 4)^2 + (y - 2)^2 = 10$ at the point C(7,1).

◉ Find gradient of line joining C and centre of circle:

$(x - 4)^2 + (y - 2)^2 = 10$
$\Rightarrow$ Circle has centre $(4,2)$
Gradient $m_1 = \frac{1-2}{7-4} = \frac{-1}{3}$

◉ Find perpendicular gradient: $m_1 = \frac{-1}{3} \Rightarrow m_2 = 3$

◉ Use $y - y_a = m(x - x_a)$ to find equation of tangent:

$m_2 = 3$, use C(7,1)
$\Rightarrow y - 1 = 3(x - 7)$
$\Rightarrow y = 3x - 20$

Introduction to Trigonometry

Applications

- Trigonometry is widely applicable to many disciplines, including engineering, navigation & geography.
- The trigonometric functions are fundamental to scientific principles involving periodic movement, such as sound or light waves, or tidal patterns.

Trigonometric ratios

The trigonometric ratios relate to the sides and angles of a right-angled triangle:

$$\sin(\theta) = \frac{o}{h}$$

$$\cos(\theta) = \frac{a}{h}$$

$$\tan(\theta) = \frac{o}{a}$$

These ratios can be used to find missing sides or angles.

▷ e.g. Find the size of x:

$$\sin(20) = \frac{x}{5}$$
$$\Rightarrow x = 5\sin(20)$$
$$\Rightarrow x = 1.71\text{cm (3sf)}$$

Inverses

The inverse functions are:

$\sin^{-1}(\theta)$ or $\arcsin(\theta)$
$\cos^{-1}(\theta)$ or $\arccos(\theta)$
$\tan^{-1}(\theta)$ or $\arctan(\theta)$

Exact value

θ	sin	cos	tan
0°	0	1	0
30°	$\frac{1}{2}$	$\frac{\sqrt{3}}{2}$	$\frac{\sqrt{3}}{3}$
45°	$\frac{\sqrt{2}}{2}$	$\frac{\sqrt{2}}{2}$	1
60°	$\frac{\sqrt{3}}{2}$	$\frac{1}{2}$	$\sqrt{3}$
90°	1	0	-

The unit circle

We can use a unit circle to define trigonometric ratios for any angle.

▶ The point C has coordinate (x,y) on a unit circle.
▶ θ is the angle that OC makes with the positive x-axis in a clockwise direction.
▶ -θ is the angle that OC makes with the positive x-axis in a anticlockwise direction.

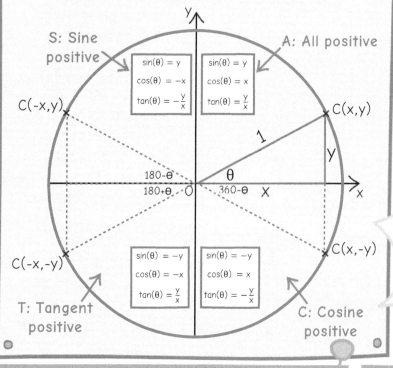

S: Sine positive

$\sin(\theta) = y$
$\cos(\theta) = -x$
$\tan(\theta) = -\frac{y}{x}$

A: All positive

$\sin(\theta) = y$
$\cos(\theta) = x$
$\tan(\theta) = \frac{y}{x}$

$C(-x,y)$ $C(x,y)$

$180-\theta$
$180+\theta$ $360-\theta$

$C(-x,-y)$

$\sin(\theta) = -y$
$\cos(\theta) = -x$
$\tan(\theta) = \frac{y}{x}$

$\sin(\theta) = -y$
$\cos(\theta) = x$
$\tan(\theta) = -\frac{y}{x}$

$C(x,-y)$

T: Tangent positive

C: Cosine positive

Periodicity and identities

Sine, cosine and tangent are periodic; the graphs of sine and cosine repeat every 360° and the graph of tangent repeats every 180°.

From this, we get the following results:

$$\sin(\theta) = \sin(\theta + 360°)$$
$$\cos(\theta) = \cos(\theta + 360°)$$
$$\tan(\theta) = \tan(\theta + 180°)$$

The graph of cosine is a translation of graph of sine by 90°: $\cos(\theta) = \sin(\theta + 90°)$

Inspection of the unit circle also gives:

$$\sin(\theta) = \sin(180° - \theta)$$
$$\cos(\theta) = \cos(360° - \theta)$$
$$\tan(\theta + 180°) = \tan(180° - \theta)$$

Area of a triangle

▷ This can be used to calculate area when two sides and the angle between them are known.

$$A_T = \tfrac{1}{2}ab\sin(C)$$

Sine rule

- The sine rule states: $\dfrac{\sin(A)}{a} = \dfrac{\sin(B)}{b} = \dfrac{\sin(C)}{c}$

- Or equivalently: $\dfrac{a}{\sin(A)} = \dfrac{b}{\sin(B)} = \dfrac{c}{\sin(C)}$

- The sine rule can be used in problems where pairs of opposite sides and angles are known.

Cosine rule

- The cosine rule states: $c^2 = a^2 + b^2 - 2ab\cos(C)$

- Or equivalently: $\cos(A) = \dfrac{b^2 + c^2 - a^2}{2bc}$

- The cosine rule can be used where all 3 sides or two sides & the angle between, are known.

Introduction to Differentiation

Applications
- Part of calculus (study of change); differentiation relates to rate of change.
- Many applications, including:
  - Rates of change in chemical reactions;
  - Links between displacement, velocity And acceleration in mechanics;
  - Finding turning points and tangents of curves.

Gradient of a curve
- The gradient of a curve at a point A is defined as the gradient of the tangent to the curve at A:

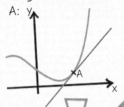

The derivative
- The derivative of the curve $y = f(x)$, denoted $f'(x)$ is defined as: $f'(x) = \lim_{h \to 0} \dfrac{f(x + h) - f(x)}{h}$

- Sometimes, different notation is used:
  - δx, representing a small change in x, is used instead of h;
  - $\dfrac{dy}{dx}$ is used instead of $f'(x)$, leading to: $\dfrac{dy}{dx} = \lim_{\delta x \to 0} \dfrac{f(x + \delta x) - f(x)}{\delta x}$
  - These differences in notation exist because calculus was developed independently by two different mathematicians (Newton and Leibniz). Both forms of notation are valid.

- The act of finding a derivative is called differentiation.
  - A function is differentiable at x if it has a derivative at x.
  - If a function is differentiable at every value of x, we say the function is differentiable.

Chord approximation
- We can calculate the gradient of the chord joining A and B. By moving B closer and closer to A, we can obtain a better approximation to the gradient of $f(x)$ at A.

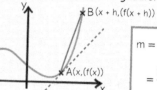

$$m = \frac{f(x + h) - f(x)}{(x + h) - x}$$
$$= \frac{f(x + h) - f(x)}{h}$$

- We imagine h becoming infinitesimally small (we say h tends to 0), leading to the definition of the gradient function $f'(x)$.

Second derivative
- If we differentiate the function $f(x)$ twice, we obtain the second derivative $f''(x)$ or $\dfrac{d^2y}{dx^2}$.
- This is understood as rate of change of the gradient of $f(x)$.
- e.g. $f(x) = 5x^3 + 3x^2$
  $\Rightarrow f'(x) = 15x^2 + 6x$
  $\Rightarrow f''(x) = 30x + 6$

From first principles
- Differentiate $f(x) = x^2$:

$$f'(x) = \lim_{h \to 0} \frac{(x + h)^2 - x^2}{h}$$
$$\Rightarrow f'(x) = \lim_{h \to 0} \frac{x^2 + 2xh + h^2 - x^2}{h}$$
$$\Rightarrow f'(x) = \lim_{h \to 0} (2x + h)$$
$$\Rightarrow f'(x) = 2x$$

- This derivative has been calculated from first principles. It would be inefficient to use this method in most cases.
- However, you do need to understand and be able to demonstrate this process for simple functions.

Polynomials
- For polynomials of the form:

$$f(x) = x^n \ (n \in \mathbb{Q})$$
$$f'(x) = \frac{df(x)}{dx} = nx^{n-1}$$

  Subtract 1 from power. Multiply by old power.

- e.g.   Constant term disappears.
$$f(x) = 4x^3 + 5x^2 - 6 - 3x^{-1}$$
$$\Rightarrow f'(x) = 12x^2 + 10x + 3x^{-2}$$

- To evaluate $f'(x)$ at x = 2:
$$f'(2) = 12(2)^2 + 10(2) + 3(2)^{-2}$$
$$\Rightarrow f'(2) = 68.75$$

Increasing and decreasing
- If $f'(x) > 0$ for every x value in the interval [a,b], then $f(x)$ is increasing on that interval.
- If $f'(x) < 0$ for every x value in the interval [a,b], then $f(x)$ is decreasing on that interval.
- e.g. $f(x) = 2x^2 - 8x$
  $\Rightarrow f'(x) = 4x - 8$
  $f(x)$ increasing for $4x - 8 \geq 0$
  $\Rightarrow f(x)$ increasing for $x \geq 2$

Other functions
- If two functions $f(x)$ and $g(x)$ are added/subtracted, differentiate each function separately:

$$\frac{d}{dx}[f(x) \pm g(x)] = \frac{d}{dx}[f(x)] \pm \frac{d}{dx}[g(x)]$$

- e.g. $f(x) = x^3 + x^2$
  $\Rightarrow f'(x) = 3x^2 + 2x$

- If a function $f(x)$ is multiplied by a constant, differentiate $f(x)$ then multiply by the constant:

$$\frac{d}{dx}[cf(x)] = c\frac{d}{dx}[f(x)]$$

- e.g. $f(x) = 5x^3$
  $\Rightarrow f'(x) = 5 \times 3x^2 = 15x^2$

Indefinite Integration

Applications

- Indefinite integration allows us to reverse differentiation to find the original function.
- This develops further to definite integration, which allows us to calculate areas under curves.

The antiderivative

- The antiderivative reverses the process of differentiation. $F(x)$ is an antiderivative of $f(x)$ if $F'(x) = f(x)$ for all x.
- e.g. $f(x) = 3x^2$ could be obtained by differentiating $F(x) = x^3$. This is one possible antiderivative for $f(x)$.
- $F(x)$ could in fact be any function in the form $\boxed{F(x) = x^3 + c}$, as the constant term disappears when we differentiate.
- c is called the constant of integration.

Integrating polynomials

For polynomials of the form:

$$f(x) = x^n \; (n \in \mathbb{Q}, \; n \neq -1)$$
$$\Rightarrow \int f(x)\, dx = \frac{x^{n+1}}{n+1} + c$$

Add 1 to power. Divide by new power.

e.g. $\boxed{f(x) = x^3 \Rightarrow \int f(x)\, dx = \dfrac{x^4}{4} + c}$

This also works for roots and other fractional powers:

$$f(x) = \sqrt[3]{x^2} = x^{\frac{2}{3}} \Rightarrow \int f(x)\, dx = \frac{3x^{\frac{5}{3}}}{5} + c$$

However, we can't integrate a power of -1 using this method, as it leads to division by 0.

Approximating area

To find the area under the curve $f(x)$ between the limits $x = a$ and $x = b$, we could approximate using rectangles with width δx:

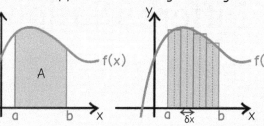

- We could obtain increasingly better approximations by using more rectangles.

- The best approximation is obtained when the width of each rectangle, δx, is infinitesimally small, i.e. when δx tends to 0.

We assume that $f(x)$ is a continuous function, and $A(x)$ is the area under the curve between a and x. We then add a small strip of width h:

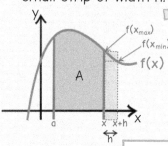

- The exact area of the strip is:

$$\boxed{A(x + h) - A(x)}$$

- There is a maximum and minimum possible value for the function $f(x)$ over the strip, and hence a maximum and minimum area:

$$\boxed{f(x_{min}) \times h \leq A(x + h) - A(x) \leq f(x_{max}) \times}$$

- Dividing by h gives: $\boxed{f(x_{min}) \leq \dfrac{A(x + h) - A(x)}{h} \leq f(x_{max})}$

- As h tends to 0: $\boxed{\begin{array}{l} x_{min} \to x \Rightarrow f(x_{min}) \to f(x) \\ x_{max} \to x \Rightarrow f(x_{max}) \to f(x) \end{array}}$, so: $\boxed{\displaystyle\lim_{h \to 0} \dfrac{A(x + h) - A(x)}{h} = f(}$

- Therefore: $\boxed{A'(x) = f(x)}$ ← The derivative of the area under $f(x)$ between a and x is equal to $f(x)$.

- Reversing the process of differentiation allows us to find $A(x)$.

Developing integral notation

We say:

The indefinite integral of $f(x)$

$$\int f(x)\, dx = F(x) + c$$

plus a constant of integration

with respect to x is equal to the antiderivative $F(x)$

We usually use the notation for "indefinite integral" rather than $F(x)$ and the word "antiderivative".

Other functions

As for differentiation, we have the following:

$$\int [f(x) \pm g(x)]\, dx = \int [f(x)]\, dx \pm \int [g(x)]\, dx$$
$$\int [cf(x)]\, dx = c\int [f(x)]\, dx$$

Remember to integrate the constant term.

e.g.
$$f(x) = 4x^3 + 6x^2 - 5 \Rightarrow \int f(x)\, dx = \frac{4x^4}{4} + \frac{6x^3}{3} - 5x + c$$
$$\Rightarrow \int f(x)\, dx = x^4 + 2x^3 - 5x + c$$

Exponential Functions

▷ An exponential function is one where the variable appears in the index.

☐ $f(x) = 2^x$ is an example of a simple exponential function.

▷ The general exponential growth function is in the form: $f(x) = ka^x, a > 0, a \neq 1$

☐ k is the initial value and a is the base or growth factor.

Applications

▷ Exponential curves are widely applicable in the sciences, and are used to model a variety of situations, including:

☐ Population growth (human/animal/bacterial);

☐ Reaction rates and radioactive decay;

☐ Compound interest and financial growth.

Growth and decay

▨ If the growth factor a > 1, the function represents exponential growth:

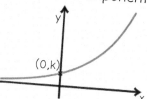

▨ If the growth factor 0 < a < 1, the function represents exponential decay:

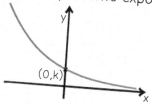

○ An exponential graph always passes through (0,k), because $a^0 = 1$.

○ $ka^x > 0$ for any values of k, a and x, so the x-axis is an asymptote.

▨ A colony of bacteria has an initial population of 100 and a growth factor of 1.4 per hour. How many bacteria will there be after 5 hours?

○ a > 1 so this situation is exponential growth.

○ Here, the growth function is $f(t) = 100 \times 1.4^t$.

○ Substitute t = 5:
$$f(t) = 100 \times 1.4^t$$
$$\Rightarrow f(5) = 100 \times 1.4^5$$
$$= 538 \text{ bacteria (3sf)}$$

▨ A car initially costs £12000, but depreciates at a rate of 15% per year. What is it worth after 10 years?

○ a = 0.85 so this situation is exponential decay.

○ Here, the growth function is $f(t) = 12000 \times 0.85^t$.

○ Substitute t = 10:
$$f(t) = 12000 \times 0.85^t$$
$$\Rightarrow f(10) = 12000 \times 0.85^{10}$$
$$= £2360 \text{ (3sf)}$$

The natural base, e

⟩ We can rewrite any exponential function using a different base.

⟩ Note that the base must be positive and can't be 1.

⟩ e.g. Base 16 can easily be changed to base 4 or 2:
$$f(x) = 16^x = 4^{2x} = 2^{4x}$$

⟩ This motivates the idea of a universal base for all exponential functions.

⟩ The most common universal base is the natural base: $e \approx 2.718281828...$

⟩ e is particularly useful because the function e^x is its own derivative, i.e. $\frac{d}{dx}(e^x) = e^x$

⟩ We also have the result: $\frac{d}{dx}(e^{kx}) = ke^{kx}$

⟩ This property of e makes it very useful when solving differential equations which represent rates of change.

Using base e

◉ A population of birds is growing exponentially and can be modelled as: $P = 200e^{0.3t}$.

How many years will it take for the population of birds to double?

Initial population = 200
Need t such that:
$400 \leq 200e^{0.3t}$
t = 2 gives P = 364.42...
t = 3 gives P = 491.92...
The population is first greater than 400 after 3 years.

Logarithms can be used to solve problems like this more efficiently than using trial and error.

Logarithms

Applications

- Logarithms are useful for solving equations involving exponential functions without using a trial and error approach.
- Historically, logarithms were used for calculations involving very large numbers - now we can use calculators.
- Logarithmic spirals appear frequently in nature; for example, in shells and the shape of galaxies like the Milky Way.

Definition

- A logarithm is the inverse of the exponential function:

  $\log_a(y)$ is the value of x such that $y = a^x$

- e.g $\log_2(16) = 4$ because $16 = 2^4$
- a is the base of the logarithm.
- $a > 0$ and $a \neq 1$.
- As we saw from graphs of the exponential function, $y > 0$.

- 10 and e (the natural base) are particularly common bases.

The natural logarithm

- We use special notation for logarithms using base e: $\log_e(y) = \ln y$.

- This function is known as the natural logarithm.
- The graph of $y = \ln(x)$ is the reflection of $y = e^x$ in the line $y = x$.

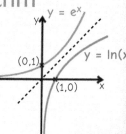

$y = e^x$
$y = \ln(x)$
$(0,1)$
$(1,0)$

Laws of logarithms

- Multiplication and division laws:

  $\log_a(x) + \log_a(y) = \log_a(xy)$

  $\log_a(x) - \log_a(y) = \log_a\left(\frac{x}{y}\right)$

- e.g: $\log_3(9) + \log_3(27) = \log_3(243)$
  $\quad\quad 2 \quad\quad\quad 3 \quad\quad\quad 5$

- Power law: $\log_a(x^k) = k\log_a(x)$

- e.g: $\log_4(8^2) = 2\log_4(8) = 3$

- We also have: $\log_a(a) = 1$
  $\quad\quad\quad\quad\quad \log_a(1) = 0$

- Logs of reciprocals (from division law):

  $\log_a\left(\frac{1}{x}\right) = -\log_a(x)$

  e.g: $\log_4\left(\frac{1}{16}\right) = -\log_4(16) = -2$

- Change of base: $\log_a(x) = \dfrac{\log_b(x)}{\log_b(a)}$

  e.g: $\log_4(16) = 2 = \dfrac{4}{2} = \dfrac{\log_2(16)}{\log_2(4)}$

Simplifying expressions

- e.g: Write as a singl logarithm in base 2:

  $3\log_2(x) + \log_2(2x) - 3$
  $= \log_2(x)^3 + \log_2(2x) - 3\log_2($
  $= \log_2(x^3)(2x) - \log_2(2)^3$
  $= \log_2\left(\frac{2x^4}{8}\right)$
  $= \log_2\left(\frac{x^4}{4}\right)$

Solving equations & inequalities

- Logarithms are used to solve equations and inequalities involving exponentials.

- e.g. Solve

  $400 \leq 200e^{0.3t}$

  (see Exponential Functions)

  $400 \leq 200e^{0.3t}$
  $\Rightarrow 2 \leq e^{0.3t}$
  $\Rightarrow \ln(2) \leq 0.3t$
  $\Rightarrow \dfrac{\ln(2)}{0.3} \leq t$
  $\Rightarrow t \geq 2.31$ (3sf)

- Some equations requ application of log law

  $5^{x-3} = 3^x$
  $\Rightarrow (x-3)\log(5) = x\log(3)$
  $\Rightarrow x\log(5) - x\log(3) = 3\log$
  $\Rightarrow x\log\left(\frac{5}{3}\right) = 3\log(5)$
  $\Rightarrow x = \dfrac{3\log(5)}{\log\left(\frac{5}{3}\right)} = 9.452$ (3dp)

  Exact value   Approximat

- Sometimes a substitut is useful:

  $3^{2x} - 4(3^x) - 12 = 0$
  $\Rightarrow (3^x)^2 - 4(3^x) - 12 = 0$
  Let $y = 3^x$:
  $\Rightarrow y^2 - 4y + 12 = 0$
  $\Rightarrow (y - 6)(y + 2) = 0$
  $\Rightarrow 3^x = 6$ and $3^x = -2$
  $\Rightarrow x = \log_3(6)$
  $\Rightarrow x = 1.631$ (3dp)

  No (real) solution

Graphs using a log scale

- We can use a logarithmic scale to plot a linear graph of an equation in the form
  $y = Ax^b$

- Taking logs of both sides gives:
  $\log_a y = \log_a(Ax^b)$ and applying log laws:
  $\log_a y = \log_a(x^b) + \log_a(A)$
  $\log_a y = b\log_a(x) + c$

  Any base can be used. Bases 10 and e are most common.
  Constant

- e.g: $y = 3x^2$

  $\log_{10} y = \log_{10}(3x^2)$
  $\log_{10} y = 2\log_{10}(x) + \log_{10}(3)$

  Gradient   y-intercept

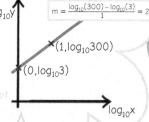

$\log_{10} y$
$m = \frac{\log_{10}(300) - \log_{10}(3)}{1} = 2$
$(1, \log_{10}300)$
$(0, \log_{10}3)$
$\log_{10} x$

Trigonometric Equations

Pythagorean identity

Applying Pythagoras' theorem to the unit circle:

$$x^2 + y^2 = 1$$
$$\Rightarrow (\cos(\theta))^2 + (\sin(\theta))^2 = 1$$
$$\Rightarrow \cos^2(\theta) + \sin^2(\theta) = 1$$

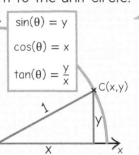

$\sin(\theta) = y$

$\cos(\theta) = x$

$\tan(\theta) = \dfrac{y}{x}$

$C(x,y)$

Powers of trigonometric functions are typically written like this.

Applications

Many problems involving periodic movement, such as travelling waves, will require solutions of trigonometric equations.

e.g. We might want to find the maximum depth of water in a tidal harbour.

The CAST diagram

Rather than draw graphs, the unit circle can be used to find solutions.

e.g. $\sin(\theta) = 0.5$ ($0° < \theta < 360°$):

● Principal value ($30°$) occurs in the A quadrant (all trig functions are positive).

● The other solution in the required interval ($150°$) occurs in the S quadrant.

● If we require solutions less than $0°$ or greater than $360°$, we add/subtract multiples of $360°$ to both solutions.

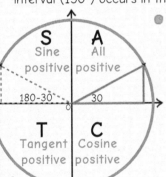

S	A
Sine positive	All positive
Tangent positive	Cosine positive
T	C

180-30   30

Solving trigonometric equations

For $y \in [-1,1]$ there are infinitely many solutions to $y = \sin(\theta)$.

e.g. Consider the equation $\boxed{\sin(\theta) = 0.5}$.

● We obtain one value (the principal value) from a calculator:

$$\boxed{\sin^{-1}(\theta) = 30°}$$

● Examination of the graph of $y = \sin(\theta)$ indicates further solutions:

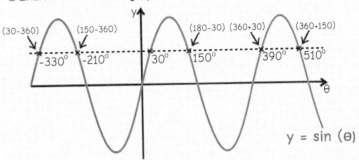

(30-360)   (150-360)   (180-30) (360+30)   (360+150)

$-330°$ $-210°$ $30°$ $150°$ $390°$ $510°$

$y = \sin(\theta)$

When solving a trigonometric equation, the required solutions are usually restricted to a given interval.

● e.g. In interval $0° < \theta < 360°$, solutions to $\sin(\theta) = 0.5$ are $30°$ & $150°$.

Using identities

Equations may need simplifying before they can be solved.

Two useful identities are:

① $\boxed{\dfrac{\sin(\theta)}{\cos(\theta)} = \tan(\theta)}$ ② $\boxed{\cos^2(\theta) + \sin^2(\theta) = 1}$

e.g. Prove: $\dfrac{\sin^4(\theta) - \cos^4(\theta)}{\cos^2(\theta)} \equiv \tan^2(\theta) - 1$

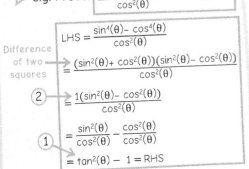

Difference of two squares

$$\text{LHS} = \frac{\sin^4(\theta) - \cos^4(\theta)}{\cos^2(\theta)}$$
$$= \frac{(\sin^2(\theta) + \cos^2(\theta))(\sin^2(\theta) - \cos^2(\theta))}{\cos^2(\theta)}$$

②

$$= \frac{1(\sin^2(\theta) - \cos^2(\theta))}{\cos^2(\theta)}$$
$$= \frac{\sin^2(\theta)}{\cos^2(\theta)} - \frac{\cos^2(\theta)}{\cos^2(\theta)}$$

①

$$= \tan^2(\theta) - 1 = \text{RHS}$$

Solving more complex equations

When solving equations, check the interval carefully and make sure you have included all solutions.

Solve $\dfrac{\sin^4(\theta) - \cos^4(\theta)}{\cos^2(\theta)} = 1$ for $0° \leq \theta \leq 360°$

$\Rightarrow$ Solve $\tan^2(\theta) - 1 = 1$ for $0° \leq \theta \leq 360°$ ← Simplify first.

$\Rightarrow \tan^2(\theta) = 2$

$\Rightarrow \tan(\theta) = \pm\sqrt{2}$ ← An easy place to lose solutions!

$\tan(\theta) = \sqrt{2}$

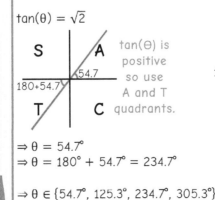

S	A
T	C

180+54.7   54.7

$\tan(\theta)$ is positive so use A and T quadrants.

$\Rightarrow \theta = 54.7°$

$\Rightarrow \theta = 180° + 54.7° = 234.7°$

$\tan(\theta) = -\sqrt{2}$

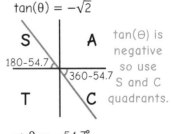

S	A
T	C

180-54.7   360-54.7

$\tan(\theta)$ is negative so use S and C quadrants.

$\Rightarrow \theta = -54.7°$

$\Rightarrow \theta = 180° - 54.7° = 125.3°$

$\Rightarrow \theta = 360° - 54.7° = 305.3°$

$\Rightarrow \theta \in \{54.7°, 125.3°, 234.7°, 305.3°\}$

Tangents, Normals & Stationary Points

Applications
- Tangents and normals are extremely useful in Mechanics/Physics, especially when working with forces.
- Finding and classifying stationary points allows us to better understand the behaviour of functions.

Definitions

- A tangent to a curve $f(x)$ at point A is a straight line touching the curve at $x = A$ with gradient equal to $f'(A)$.
- A normal is perpendicular to the tangent.

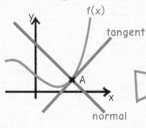

- The product of these gradients is -1.

- The point A is a stationary point if $f'(A) = 0$.
- The tangent at A is horizontal.
- There are three types of stationary point:

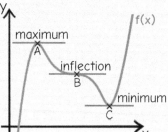

- Maxima and minima are local - there may be other points where $f(x)$ has a greater or smaller value.

Equation of tangent

▷ e.g: Find the equation of the tangent to $f(x) = 2x^2 - 4x + 5$ at $x = 3$:
○ Use differentiation to find the gradient of the tangent at $x = 3$:
$$\Rightarrow f'(x) = 4x - 4$$
$$\Rightarrow f'(3) = 8$$

○ Find value of y at $x = 3$: $f(3) = 11$
Hence A = (3,11).

○ Apply $y - y_a = m(x - x_a)$:
$$\Rightarrow y - 11 = 8(x - 3)$$
$$\Rightarrow y = 8x - 13$$

Equation of normal

▷ e.g: Find the equation of the normal to $f(x) = 2x^2 - 4x +$ at $x = 3$:
○ Find gradient of normal. Use the fc that the product of perpendicular gradients is -1:

$f'(3) = 8$
$\Rightarrow$ gradient of tangent = 8
$\Rightarrow$ gradient of normal = $-\frac{1}{8}$

○ Apply $y - y_a = m(x - x_a)$:
$$\Rightarrow y - 11 = -\frac{1}{8}(x - 3)$$
$$\Rightarrow 8y + x = 91$$

Finding stationary points

- To find the stationary points of a function $f(x)$, we solve $f'(x) = 0$.

▷ e.g. Find the stationary points of $f(x) = x^4 + 2x^3$:

$f(x) = x^4 + 2x^3$
$\Rightarrow f'(x) = 4x^3 + 6x^2$

$4x^3 + 6x^2 = 0$
$\Rightarrow 2x^2(2x + 3) = 0$
$\Rightarrow x = 0$ and $x = -\frac{3}{2}$

$A = \left(-\frac{3}{2}, -\frac{27}{16}\right)$
$B = (0,0)$

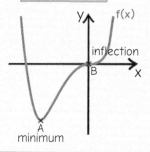

Second derivative

○ Use the second derivative to classify stationary points.

Minimum $f''(x) > 0$ Maximum $f''(x) < 0$

○ If $f''(x) = 0$, the point could be any of the three types of point. To be an inflection, $f'(0)$ must have the same sign either side of the point.

Classifying stationary points

- e.g. For $f(x) = x^4 + 2x^3$, we have $f''(x) = 12x^2 + 12x$.

▷ Point A:
$x = -\frac{3}{2} \Rightarrow f''(x) = 9$
$f''(x) > 0$ (minimum)

▷ As we move through point A, $f'(x)$ goes from negative to positive. (Note that we don't **need** to check this - $f''(x) > 0$ is sufficient to identify a minimum).

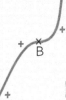

Note $f'(B) = 0$

▷ Point B:
$x = 0 \Rightarrow f''(x) = 0$

▷ Consider values of $f'(x)$ either side of (0,0):
$x = -0.1 \Rightarrow f'(x) = 0.056$
$x = 0.1 \Rightarrow f'(x) = 0.064$

▷ As we move through point B, $f'(x)$ remains non-negative, so B mus be an inflection.

Definite Integration

The definite integral

A definite integral is evaluated between specified x limits. Each definite integral represents an area A bounded by a curve f(x), the limits x = a and x = b and the x-axis.

$$\int_a^b f(x)\,dx = \Big[F(x)\Big]_a^b = F(b) - F(a)$$

The definite integral of f(x) between x = a and x = b | is equal to | the difference between the antiderivative F(x) evaluated at x = a and x = b.

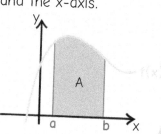

Applications

- Definite integration allows us to find the area under curves and volumes of revolution of shapes around axes.
- This has many applications in engineering, including modelling airfoils and sails.
- We could use integration to find the area under a velocity-time graph (displacement).

The Fundamental Theorem of Calculus

The FTC formally links differentiation and definite integration:

If the function f is continuous on [a,b] and the function F is defined as:

$$F(x) = \int_a^x f(t)\,dt \quad \text{for all x in } [a,b]...$$

...then F is differentiable in [a,b] and F'(x) = f(x), i.e.

$$\frac{d}{dx}\int_a^x f(t)\,dt = f(x) \quad \text{for all x in } [a,b].$$

Evaluating definite integrals

To evaluate a definite integral,

e.g. $\displaystyle\int_2^4 3x^2\,dx$

Find the antiderivative F(x):

$$F(x) = \int 3x^2\,dx = \frac{3x^3}{3} + c$$

Substitute in the limits to evaluate the difference between F(b) and F(a):

$$\int_2^4 3x^2\,dx = \left[\frac{3x^3}{3}\right]_2^4 = [x^3]_2^4 = (4^3) - (2^3) = 56$$

The area under a curve

e.g. Find the area of the region R:

$$f(x) = x^3 + 2x^2$$

$$\int_{-2}^0 x^3 + 2x^2\,dx = \left[\frac{x^4}{4} + \frac{2x^3}{3}\right]_{-2}^0$$

$$= (0) - \left(\frac{-4}{3}\right) = \frac{4}{3}$$

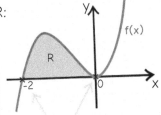

Integrate between x = 0 and x = -2.

Curves above and below the axis

e.g. Find the area between the curve f(x) and the x axis:

- If we integrate f(x) over the entire region between 0 and 3, area B will have a negative value and incorrectly "cancel out" part of area A.
- To avoid this, treat each region separately:

$f(x) = x^3 - 5x^2 + 6x$

$$A = \int_0^2 x^3 - 5x^2 + 6x\,dx = \left[\frac{x^4}{4} - \frac{5x^3}{3} + 3x^2\right]_0^2 = \frac{8}{3}$$

$$B = \int_2^3 x^3 - 5x^2 + 6x\,dx = -\frac{5}{12}$$

Total area $= A + B = \dfrac{8}{3} + \dfrac{5}{12} = 3\dfrac{1}{12}$

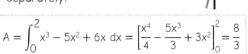
Treat area B as positive.

Non-smooth functions

- A non-smooth function has derivatives which are not continuous.
- For example, functions involving modulus can be non-smooth.
- To find the area of the shaded region, treat each part separately:

$$f(x) = |2x| - x$$

$$|2x| = \begin{cases} 2x, & x \geq 0 \\ -2x, & x < 0 \end{cases}$$

A: $f(x) = -2x - x = -3x$

$$A = \int_{-1}^0 -3x\,dx = \left[\frac{-3x^2}{2}\right]_{-1}^0 = -\frac{3}{2}$$

B: $f(x) = 2x - x = x$

$$B = \int_0^2 x\,dx = \left[\frac{x^2}{2}\right]_0^2 = 2$$

Total area $= A + B = \dfrac{7}{2}$

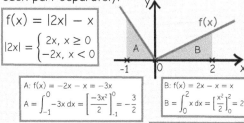

- Note that we could also have done this using area of triangles.

Population & Samples

In a statistical study, the population is the set of things that we collect data from. This may be as large a population as the inhabitants of the UK, or smaller populations such as students in a year group.

> Population parameters (such as mean/variance) describe the entire population. For large populations, we calculate estimates for these from samples.

A sample is a portion of the population, which is (hopefully) selected in such a way that it represents the entire population. There is a variety of methods for doing this.

Applications

When working with statistics, it is often impractical to collect data from an entire population.

Sampling is used to select a smaller amount of data that is representative of the wider population.

Sampling methods

> Simple random sampling:
- Each member of the population has the same probability of being chosen, but selections are not independent.
- One method is to number the members of a population, then use a random number generator to select a certain amount.

> Systematic sampling:
- Each choice made independently and each member of the population has the same probability of being chosen.
- One method is to number the members of a population, then use a random number generator to select a certain amount.

> Both methods are useful if we have little information about the population, but they require access to data about every member of the population.

> Stratified sampling:
- Selects proportions of various categories in a population (e.g. age/gender).
- An effective sampling method, but requires detailed knowledge of initial population.

> These sampling methods may not be as effective, as they may not accurately reflect the population:

> Opportunity sampling:
- Collects from members of the population for which data is available.

> Quota sampling:
- Selects a certain number of members from each category (e.g. 20 people between 21-34).

> Cluster sampling:
- Selects groups designed to represent the population, then chooses a random sample from these.

Statistics and estimators

A statistic is a random variable which is a function of a sample (e.g. sample mean).

An estimator is a statistic used to estimate a parameter (e.g. calculating the sample mean to estimate the mean of the population).

We aim to select a sample and statistics that will give good estimates of the parameters.

Biased & unbiased estimators

The sample statistic R is an unbiased estimator for population parameter ρ if:

$$\mathbb{E}(R) = \rho$$

The sample mean is an unbiased estimator of the population mean.

The sample variance is a biased estimator of the population variance - the sample variance tends to be lower, because we are less likely to select extreme values in a sample.

Biased and unbiased variance

Sample variance calculated as:

$$s^2 = \frac{1}{N}\sum_{i=1}^{N}(X_i - \bar{X})^2$$

is a biased estimator.

If we divide by N-1 rather than N, we obtain an unbiased estimator:

This works because there are only N-1 independent residuals (difference between data point and mean).

$$s^2 = \frac{1}{N-1}\sum_{i=1}^{N}(X_i - \bar{X})^2$$

Presenting and Interpreting Data

Applications
- All of the statistical tools detailed here are used to transform raw data into a more manageable form to allow us to draw conclusions about the original population.
- Charts and graphs are used to display data in visual formats. Statistics such as mean and variance can be used to summarise sample data.

Averages

- An average is a measure of central tendency; we try to find a representative value for the data set.

- Let $\{X_1, X_2, X_3, ..., X_N\}$ be an ordered data set with N observations.

- The midpoint can be useful when dealing with ordered data, but is affected by outliers:

$$\text{midpoint} = \frac{1}{2}(X_1 + X_N)$$

- The mode is most useful for qualitative data or data with many repeats. It is otherwise not particularly representative:

$$\text{mode} = \text{most frequent value}$$

- The median provides a good measure of the "middle" of the data. It is particularly useful when the data is skewed or contains a few outliers:

$$\text{median} = \begin{cases} X_n & \text{for odd } N = 2n - 1 \\ \frac{1}{2}(X_n + X_{n+1}) & \text{for even } N = 2n \end{cases}$$

- The mean also provides a generally representative average, but is affected by extreme values, so is less useful with skewed data sets:

$$\text{mean} = \frac{1}{N}\sum_{i=1}^{N} X_i$$

Interpolation

- Interpolation is a technique used with grouped data to obtain a better estimate for the median or quartiles.

  - Interpolation can be carried out quite simply using a cumulative frequency graph:
  - Alternatively, find the median class, then the following can be used:

$$\text{median} = \text{lower class boundary} + \text{class width} \times \text{proportion through group}$$

e.g. For height data for a group of 70 students:

Height h (cm)	Frequency
$140 \le h < 150$	11
$150 \le h < 160$	20
$160 \le h < 165$	19
$165 \le h < 170$	15
$170 \le h < 180$	5

N = 70 so median is 35.5th value. This lies in class $160 \le h < 165$.

160 — x — 165
31   35.5   50

$$\text{Median} = 160 + 5 \times \frac{4.5}{19} = 161.2\text{cm (4sf)}$$

- This can be adapted for quartiles or other required values.

Variance & standard deviation

- Variance is a measure of how far a set of data is spread out from the mean.

$$\text{Variance} = \sigma^2 = \frac{1}{N}\sum_{i=1}^{N}(X_i - \mu)^2 = \frac{1}{N}\sum_{i=1}^{N}X_i^2 - \mu^2$$

This version is often easier to work with.

- A simple way to think about this is "the mean of the squares minus the square of the mean".

- Standard deviation is the square root of the variance, and is more intuitive as a measure of spread, as it has the same units as the original data (variance has squared units).

- e.g. Estimates for the mean and standard deviation for the heights of pupils:

Height h (cm)	f	m	fm	fm²
$140 \le h < 150$	11	145	1595	231275
$150 \le h < 160$	20	155	3100	480500
$160 \le h < 165$	19	162.5	3087.5	501718.8
$165 \le h < 170$	15	167.5	2512.5	420843.8
$170 \le h < 180$	5	175	875	153125
			11170	1787463

$$\mu = \frac{11170}{70} = 159.57\text{cm (2dp)}$$

$$\sigma = \sqrt{\frac{1787463}{70} - \left(\frac{11170}{70}\right)^2} = 8.49\text{cm (2dp)}$$

It is often not necessary to compute these in full; you can also use the statistics functions on your calculator.

Coded data

- If data values are large, we can code the data to make it easier to work with.

- For example, we could code the height data above by subtracting 140 from each data value and then dividing by 5:

$$H = \frac{h - 140}{5}$$

H	f
$0 \le H < 2$	11
$2 \le H < 4$	20
$4 \le H < 5$	19
$5 \le H < 6$	15
$6 \le H < 8$	5

Coded data is dimensionless

- For coding

$$y = \frac{x - a}{b}$$

we have:

$$\mu_y = \frac{\mu_x - a}{b} \text{ so } \mu_x = b\mu_y + a$$

$$\sigma_y = \frac{\sigma_x}{b} \text{ so } \sigma_x = b\sigma_y$$

The spread of the data is not affected by adding or subtracting, as this has the effect of translating data points and does not change the shape of the distribution.

Graphs and charts

The following graphs and charts can be used:

▷ Bar charts for discrete data - include gaps between bars and all bars are the same width.

▷ Dot plots for small data sets - each data point is plotted as an individual dot. The number of dots in a given column indicates the frequency.

▷ Stem and leaf diagram - this gives a good representation of the shape of the data and retains all the detail.

▷ Histogram - use for large sets of grouped (usually continuous) data. With a histogram, the area of each bar represents the scaled frequency of the group.

▷ Cumulative frequency graph - useful for analysing the shape of large data sets. A cumulative frequency curve can be used to interpolate estimates for median & quartiles.

▷ Box plot - displays median and quartiles for a data set, useful for comparing sets of data.

Bar charts

▷ Bar charts are not suitable for continuous data, but are very useful for discrete and qualitative data.

▷ For a simple data set, a bar chart gives a clear picture of the distribution, and the mode and range can easily be identified.

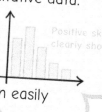
Positive skew clearly shown

Skew

Data is skewed if its spread is not symmetrical.

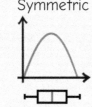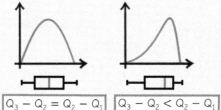

Positive skew	Symmetric	Negative skew

| $Q_3 - Q_2 > Q_2 - Q_1$ | $Q_3 - Q_2 = Q_2 - Q_1$ | $Q_3 - Q_2 < Q_2 - Q_1$ |

Skew coefficient 1: $\boxed{\text{skew} = \dfrac{Q_3 - 2Q_2 + Q_1}{Q_3 - Q_1}}$

○ This returns a number between +1 (positive skew) and -1 (negative skew).

○ A skew of zero does not necessarily mean that the data is symmetric - consider the shape of the data when interpreting the skewness coefficient.

Skew coefficient 2: $\boxed{\text{skew} = \dfrac{3(\mu - m)}{\sigma}}$

○ μ is the mean, m is the median and σ is the standard deviation.

○ e.g. Using the height data from the previous page with the estimated mean and standard deviation, and interpolated median:

$$\mu = 159.27, \quad m = 161.2, \quad \sigma = 8.49$$

$$\text{skew} = \frac{3(\mu - m)}{\sigma} = \frac{3(159.27 - 161.2)}{8.49}$$

$$\Rightarrow \text{skew} = -0.682$$

Skew coefficient 3: $\boxed{\text{skew} = \dfrac{(\mu - M)}{\sigma}}$

○ μ is the mean, M is the mode and σ is the standard deviation.

○ This skew coefficient is appropriate to use if the range of data is discrete rather than continuous.

Histograms

⬤ The area of each bar is proportional to frequency:

$$\text{frequency density} = \frac{\text{frequency}}{\text{class width}}$$

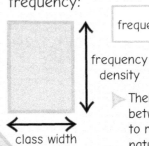
frequency density

class width

▷ There should be no gaps between bars of a histogram to represent the continuous nature of the data set.

⬤ A histogram for the height data is as follows:

Height h (cm)	f	Class width	Frequency density
$140 \leq h < 150$	11	10	1.1
$150 \leq h < 160$	20	10	2
$160 \leq h < 165$	19	5	3.8
$165 \leq h < 170$	15	5	3
$170 \leq h < 180$	5	10	0.5

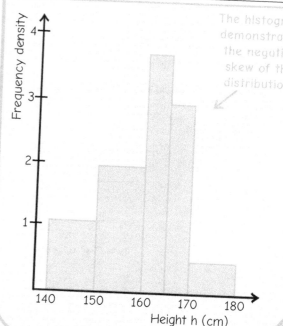

The histogram demonstrates the negative skew of the distribution.

Dot plots

- In addition to bar charts, dot plots are also used with small discrete data sets.

- For very small sets of data, a dot plot can be easier to read - you can immediately see the frequency of each category.

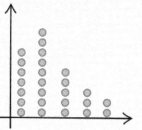

- One advantage of a dot plot over a bar chart for small sets of discrete data is that dots can be labelled or coloured to communicate additional information. Dots could be replaced with small pictures, making a visually attractive chart for publication.

Box plots and quartiles

- Quartiles divide the data into four equal parts:
  - Lower quartile = Q_1
  - Median = Q_2
  - Upper quartile = Q_3

  Q_1 is $\frac{1}{4}$ of the way through the data and Q_3 is $\frac{3}{4}$ of the way through the data. The quartiles can be thought of as the medians of each half of the data.

- The interquartile range, represented by the length of the box, is $Q_3 - Q_1$.

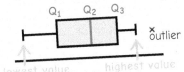

- Sometimes, data may be broken into other groupings, such as percentiles, as we may be interested in the top or bottom 10% of the data set.

Cumulative frequency graphs

A cumulative frequency graph allows us to analyse large data sets quickly.

- e.g. For the height data:

140cm is the lowest class boundary so the curve begins here.

Height h (cm)	f	Cumulative frequency
$140 \leq h < 150$	11	11
$150 \leq h < 160$	20	31
$160 \leq h < 165$	19	50
$165 \leq h < 170$	15	65
$170 \leq h < 180$	5	70

Data points are plotted using the endpoints of each class.

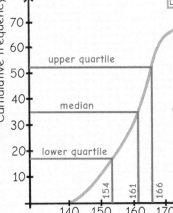

Points can be joined with a smooth curve or with line segments between points (piecewise linear).

We can estimate the median and quartiles (or other required percentiles) using the graph.

- These are rough estimates; often sufficient for large data sets, but interpolation can give a more accurate estimate.

- It is then very straightforward to draw a box plot for this data, which allows us to more easily identify the shape of the distribution.

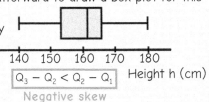

$Q_3 - Q_2 < Q_2 - Q_1$

Negative skew

- Using skew coefficient 1 gives:

$$\text{skew} = \frac{166 - 2 \times 161 + 154}{166 - 154} = -0.167 \text{ (3sf)}$$

Outliers

- There are different definitions of outliers (points which lie sufficiently far away from the average).
  - We usually only extend the whiskers on a box plot to 1.5 x the length of the box; any points outside the whiskers are plotted as single points and considered outliers.
  - The data point X is an outlier if:

  $$X < Q_1 - \frac{3}{2}(Q_3 - Q_1) \text{ or } X > Q_3 + \frac{3}{2}(Q_3 - Q_1)$$

- For the height data:

  $$X < 154 - \frac{3}{2}(166 - 154) = 136$$

  $$X > 166 + \frac{3}{2}(166 - 154) = 184$$

  Less than 136cm would be an outlier.

  Greater than 184cm would be an outlier.

As the lowest possible value is 140cm and the greatest is 180cm, there are no outliers in the data (using this definition).

Median or mean?

- The median and interquartile range are **robust** statistics; they are not affected as much by extreme values or outliers.

- The mean and standard deviation are affected significantly more by outliers; they are not robust statistical measures.

- The mean and standard deviation are only appropriate when the distribution is fairly symmetrical and does not contain many outliers.

Presenting Multivariate Data

Applications
- Multivariate data contains information about more than one variable.
- Techniques described here allow us to look for relationships or connections between variables.
- For example, we may want to consider whether sugar intake is linked to heart disease.

Scatter graphs & correlation

Bivariate data is data with two variables. Scatter graphs are used to represent a set of such data.

- The word "correlation" is used to describe the relationship between the two variables:

Positive correlation
As one variable increases, the other increases.

Negative correlation
As one variable increases, the other increases.

- If there is no relationship, we say there is no correlation.
- Correlation can be strong (points clustered close to line of best fit) or weak (points further away).

Correlation & causation

- Correlation between two variables does not imply causation.
- Statistical analysis can never prove that two variables are causally linked, so we must be careful when drawing conclusions.

Covariance

Covariance works on the principle of translating a given set of bivariate data to a standardised plot centred around the origin.

- This is done by coding both sets of data by subtracting the relevant mean from each data point: $(X_i - \mu_x, \ Y_i - \mu_y)$

These are the coded versions of the data sets shown to the left

Positive correlation; points tend to lie in quadrants 1 and 3.

Negative correlation; points tend to lie in quadrants 2 and 4.

$(X_i - \mu_x)(Y_i - \mu_y) > 0$ $(X_i - \mu_x)(Y_i - \mu_y) < 0$

Summing these products motivates the definition of covariance:

$$\text{cov}(X,Y) = \sigma_{xy}$$
$$= \frac{1}{N}\sum_{i=1}^{N}(X_i - \mu_x)(Y_i - \mu_y) = \frac{1}{N}\sum_{i=1}^{N}(X_i Y_i - \mu_x \mu_y)$$

- Positive covariance implies positive correlation; negative covariance implies negative correlation and covariance close to zero implies no correlation.

Pearson's correlation coefficient

Covariance (like variance) is affected by scaling.

- We can obtain a constant measure for correlation by dividing covariance by the standard deviation of each variable
- This is Pearson's correlation coefficient, sometimes also called the product-moment correlation coefficient:

$$\rho_{XY} = \frac{\sigma_{XY}}{\sigma_X \sigma_Y}$$

← covariance of X and Y
← standard deviations of X and Y

e.g. Calculate Pearson's coefficient for the following data on height and weight of ten adults:

Height (x cm)	Weight (y kg)
170	65
165	61
154	62
162	70
148	52
169	67
172	70
180	75
182	79
176	68

$$\sigma_{XY} = \frac{\Sigma xy}{N} - \mu_x \mu_y = \frac{112908}{10} - 167.8 \times 66.9 = 64.9$$

$\sigma_x = \sqrt{106.56}$
$\sigma_y = \sqrt{51.69}$

Use exact values to avoid rounding errors.

$$\rho_{XY} = \frac{\sigma_{XY}}{\sigma_X \sigma_Y} = \frac{64.98}{\sqrt{106.56}\times\sqrt{51.69}} = 0.8755 \text{ (4dp)}$$

- A coefficient of 0.8755 indicates some positive correlation.

Regression lines

Rather than plot a line of best fit by eye, we can more formally calculate a regression line for a set of bivariate data:

$$Y = mX + c \text{ with } m = \frac{\sigma_{XY}}{\sigma_X^2}, \ c = \mu_y - m\mu_x.$$

For the height/weight data:

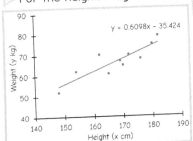

y = 0.6098x − 35.424

This line is used to interpolate values within the range of the data set.

e.g. Height 160cm predicts weight 62.1kg (3sf).

Kinematics in One Dimension

- Kinematics is the mathematical description of movement - i.e. an object's position, velocity and acceleration.
- For example, we might model a car's journey on a motorway and use this to position speed cameras.

Vector variables

In kinematics, direction of motion is important; many variables are vectors:

- We use the vector quantity displacement, **s**, rather than distance. Displacement is the position of an object relative to its starting position at t = 0.
- We use the vector quantity velocity, **u**, **v**, rather than speed. **u** is the initial velocity at t = 0; **v** is the velocity at a given time t.
- Acceleration, **a**, is also a vector quantity.
- Positive/negative quantities indicate direction of motion.
- **r** refers to the position of an object; r_0 is used for the initial position at t = 0.

SUVAT equations of motion

These equations only apply in situations with constant acceleration.

$$v = u + at$$
$$s = \frac{u+v}{2}t$$
$$s = ut + \frac{1}{2}at^2$$
$$s = vt - \frac{1}{2}at^2$$
$$v^2 = u^2 + 2as$$

- In practice, many objects don't move with constant acceleration, but it is often appropriate to model them as doing so.
- e.g. A car is travelling along the motorway. At t = 0, the car passes a camera that records a velocity of 38m/s. 3 minutes later, the car has travelled 5220m and its velocity is recorded by another camera.

a) Find and interpret the constant acceleration of the car.

Cam 1	s = 5220m	Cam 2
t = 0s		t = 180s
u = 38m/s		v = ?

Always draw a diagram!

$$s = ut + \frac{1}{2}at^2$$
$$\Rightarrow 5220 = 38 \times 180 + \frac{1}{2}(180)^2 \times a$$
$$\Rightarrow a = \frac{2(5220 - 38 \times 180)}{180^2}$$
$$\Rightarrow a = -0.1m/s^2$$

a is negative, so the car is decelerating at 0.1m/s².

b) A speeding ticket is issued if the average speed between the two cameras is greater than 30m/s. Will the driver receive a ticket?

$$s = \frac{38+v}{2}t$$
$$\Rightarrow 5220 = \frac{38+v}{2} \times 180$$
$$\Rightarrow v = \frac{5220 \times 2}{180} - 38 = 20m/s$$

Average velocity $= \frac{38+20}{2}$
$= 29m/s$

The driver will not receive a ticket.

Links to calculus

Velocity and acceleration are linked to displacement as follows:

- Velocity is the rate of change of displacement. $v(t) = \frac{ds}{dt}(t)$ — *Written as functions of time (t)*
- Acceleration is the rate of change of velocity and is therefore the second derivative of displacement. $a(t) = \frac{dv}{dt}(t) = \frac{d^2s}{dt^2}(t)$

- e.g. A car is moving with non-constant acceleration. Its velocity is: $v(t) = \frac{t^2}{20}$ Find its acceleration and displacement at t = 20s.

$$a(t) = \frac{dv}{dt}\left(\frac{t^2}{20}\right) = \frac{t}{10} \Rightarrow a(20) = \frac{20}{10} = 2m/s^2$$

$$s(t) = \int_0^{20} \frac{t^2}{20} dt = \left[\frac{t^3}{60}\right]_0^{20} = 133m \text{ (3sf)}$$

Non-constant acceleration

For non-constant acceleration, the following apply:

- Displacement:
$$s(t_1) = \int_0^{t_1} v(t) \, dt$$

- Velocity:
$$v(t_1) = v(0) + \int_0^{t_1} a(t) \, dt$$

Displacement-time graphs

The gradient of the tangent to a displacement-time graph is velocity.

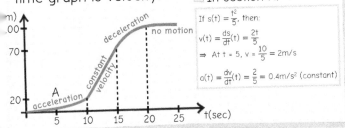

In section A:

If $s(t) = \frac{t^2}{5}$, then:
$$v(t) = \frac{ds}{dt}(t) = \frac{2t}{5}$$
$$\Rightarrow \text{At } t = 5, v = \frac{10}{5} = 2m/s$$
$$a(t) = \frac{dv}{dt}(t) = \frac{2}{5} = 0.4m/s^2 \text{ (constant)}$$

Velocity-time graphs

The gradient of the tangent to a velocity-time graph is acceleration.

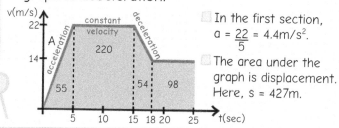

- In the first section, $a = \frac{22}{5} = 4.4m/s^2$.
- The area under the graph is displacement. Here, s = 427m.

Introduction to Forces

A force is a push or pull upon an object that arises from the interaction between two objects.

Weight is the force exerted on a force due to gravity.

A normal reaction force opposes weight when an object is placed on a surface.

Resistance is any force opposing motion, such as friction.

Tension is a pulling force transmitted by a rope (or similar)

Objects may move under driving or pushing forces.

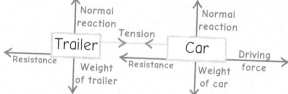

Force is a vector; it has magnitude and direction.

Force is measured in Newtons (N).

Applications

Every object, whether in motion or stationary, is subject to forces such as gravity or friction.

Engineers apply understanding of forces to ensure that planes fly as expected, or that constructions are built safely.

Resultant forces

A resultant force is the sum of all forces acting on an object in a certain direction.

e.g. A system of a car towing a trailer is subject to the forces shown:

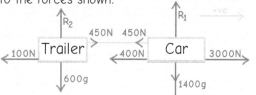

Horizontally we have:

$F(\rightarrow) = 3000 + 450 = 3450N$

$F(\leftarrow) = 400 + 450 + 100 = 950N$

Resultant $F = 3450 - 950 = 2500N$

By Newton's Second Law:

$F = ma \Rightarrow 2500 = (1400 + 600)a$

$\Rightarrow a = \frac{2500}{2000} = 1.25m/s^2$ Mass of car and trailer

Newton's Laws of Motion

First Law: An object remains in its state of motion unless acted upon by an external force.

- If an object is moving at constant speed in a straight line will continue to do so unless acted on by another force.
- If an object at rest, it will remain so unless acted on by another force.
- This means that an object does **not** have to have a force acting upon it to be in motion.

Second Law: The acceleration of an object with constant mass is directly proportional to the magnitude of the force acting upon it.

- This is more commonly written: $\boxed{F = ma}$.
- This law allows us to link kinematics and forces.
- Acceleration due to gravity is $g = 9.81m/s^2$ (3sf). Therefore weight can be written: $\boxed{W = mg}$

Equilibrium

A object is in equilibrium if and only if the resultant force is zero.

An object that is in equilibrium has constant (or zero) velocity. Its acceleration is zero.

By Newton's First Law, this does not mean that the object is at rest; it is only at rest if its velocity is zero.

e.g. Consider the car only. Vertically, the car has zero acceleration (and in this case, zero velocity). Therefore the vertical components are in equilibrium, so the resultant force is zero:

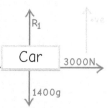

$R_1 - 1400g = 0 \Rightarrow R_1 = 1400g$

Using $g = 9.8m/s^2$, $R_1 = 13720N$

Splitting into component

When dealing with forces that are not simply horizontal or vertical, it is useful to split them into their horizontal and vertical components.

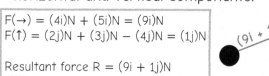

$F(\rightarrow) = (4i)N + (5i)N = (9i)N$

$F(\uparrow) = (2j)N + (3j)N - (4j)N = (1j)N$

Resultant force $R = (9i + 1j)N$

Magnitude of $R = |R| = \sqrt{9^2 + 1^2} = \sqrt{82} = 9.06N$ (3s

Direction (angle Θ) of R given by $\tan(\Theta) = \frac{1}{9}$

$\Rightarrow \Theta = 6.34°$ (3sf)

Forces to Model the Real World

Applications

- Situations in the real world are often complicated and may involve resistance forces, such as friction or air resistance.
- When creating a model for a real world situation, we need to take these forces into account.

Newton's Third Law of Motion

This states that all forces in the Universe occur in equal but oppositely directed pairs.

▷ One force in the pair is typically called the action force and the other is the reaction force.

One frequently used example of this is the normal reaction force exerted on an object by a surface that is in contact with the object.

R
W = mg

▷ The normal reaction force acts in a direction perpendicular to the surface. On a horizontal surface, the normal reaction force is equal to the weight of the object.

▷ We typically use R to denote the normal reaction force.

▷ e.g. If m = 5kg and g = $9.8m/s^2$, we obtain R = W = 49N.

Modelling assumptions

▷ When modelling a real-life situation, we can simplify some aspects of the problem as follows:

☐ A smooth surface is frictionless.

☐ If an object is in a vacuum, there is no air resistance.

☐ If a connector such as a rope or rod is described as light, we can ignore the weight of the connector.

☐ If a connector is described as inelastic, it cannot stretch or compress, so the magnitude of the tension force is the same at every point.

☐ If an object is described as uniform, we assume that there are no variations in properties such as the object's density.

☐ If an object is described as a particle, we model all forces as acting on a single point on the object. We always make this assumption in A-level Mathematics.

Thrust and resistance forces

Thrust is an example of a driving force with an equal and opposite force pair.

Air pushed through engine ← ● Plane moves forward →

▷ e.g. An aeroplane generates thrust by pushing air in the direction opposite to flight.

A resistance force opposes motion, and always acts in the opposite direction from (attempted) motion.

▷ Friction is a resistance force between two surfaces; it occurs when one surface is attempting to slide against the other.

Normal reaction ↑   a →
Friction ←   ●   → Force causing motion
Weight = mg ↓

If m = 10kg, a = $1.5m/s^2$ and F_r = 30N, what is the magnitude of the force P causing motion?
Resultant F = ma = 10×1.5 = 15N
P = F + F_r = 30 + 15 = 45N

▷ Air resistance is caused by friction between the air and an object.

Internal and external

○ Internal forces act between objects forming part of a system (e.g. components inside a car's engine).
○ External forces act on a system from outside the system (e.g. the driving force or friction acting on the car).

Connected particles

▷ e.g. Find the tension in the rope and acceleration of the system:

a ←   R ↑
T m₁ 4kg Fr = 0.5N
T ↑
a ↓ m₂ 8kg ↓4g
↓8g

Resultant F_1 = T − Fr
Resultant F_2 = 8g − T

Applying F = ma gives F_1 = 4a and F_2 = 8a

Using g = $10m/s^2$ and equating expressions for F_1 and F_2:
4a = T − 0.5 ⇒ 8a = 2T − 1
8a = 80 − T

Solving simultaneously gives:
2T − 1 = 80 − T ⇒ T = 27N
8a = 80 − 27 ⇒ a = $6.625m/s^2$

▷ Before beginning a problem, look at any modelling assumptions.
▷ Try to determine the likely motion of the system.
▷ Draw a diagram showing all forces.
▷ If particles are connected and have no relative motion, they can be treated as a single particle.

Introduction to Probability

Definitions

- An experiment is a procedure which can be repeated and has defined outcomes (possible results).
- A sample space is the collection of all possible outcomes
- An event is a subset of the sample space and is a collection of outcomes from the experiment.
- Events/outcomes are called equally likely if they have the same theoretical probability of happening.
- For an event to be chosen at random, all events must be equally likely.

Applications

- Probability is used by insurance companies to set the price of premiums.
- It is used to calculate odds of winning or losing chance games such as the Lottery.
- It is also applied in the medical profession to decide whether or not to screen for diseases or to choose a treatment plan.

▷ Venn diagrams are useful when solving probability problems.

- e.g. Given:
  P(A) = 0.3
  P(B) = 0.2
  P(A∪B) = 0.35,
  find P(A∩B').

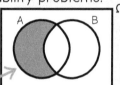

P(A∪B) = P(A) + P(B) − P(A∩B)
⇒ 0.35 = 0.3 + 0.2 − P(A∩B)
⇒ P(A∩B) = 0.15

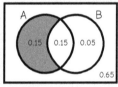

⇒ P(A∩B') = 0.15

Notation and Venn diagrams

▷ The probability of event A occurring is written P(A).

- e.g. Consider rolling a fair six-sided die, so Ω = {1,2,3,4,5,6}.
- If A is the event "rolling an even number", then $P(A) = \frac{1}{2}$.

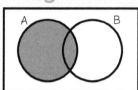

▷ The probability of event A not occurring is written P(A'). Here A' is called the complement of A.

- Here, P(A') is also $\frac{1}{2}$.
- $P(A') = 1 - P(A)$

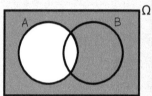

▷ The intersection of A and B, written A ∩ B, is the event that both A and B occur.

- If B is the event "rolling a prime number", then $P(A∩B) = \frac{1}{6}$.

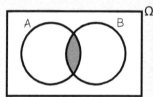

▷ The union of A and B, written A ∪ B, is the event that either A or B (or both) occur.

- $P(A∪B) = \frac{2}{3}$.

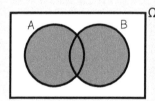

▷ The addition rule links probabilities of the union and intersection: $P(A∪B) = P(A) + P(B) - P(A∩B)$

Types of event

- Events A and B are mutually exclusive if only one of these events can occur at the same time.

  - If A and B are mutually exclusive, then the intersection is empty:
    P(A∩B) = 0
    ⇔P(A∪B) = P(A) + P(B)

- Events A and B are independent if A happening has no influence upon the probability of B happening.

  Events are independent
  ⇔ P(A∩B) = P(A) × P(B)

  - We can use this result to show that two events are independent.

Tree diagrams

- Tree diagrams can be used to represent probabilities of two events happening in succession.

  ▷ e.g. There are 3 red and 4 blue balls in a bag. One is picked and set aside, then another is picked. Find the probability that the second ball is blue.

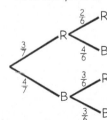

$P(RB) = \frac{3}{7} \times \frac{4}{6} = \frac{12}{42}$

$P(BB) = \frac{4}{7} \times \frac{3}{6} = \frac{12}{42}$

The probability that the second ball is blue is $\frac{4}{7}$

Discrete Probability Distributions

Applications

Probability distributions are used to model the expected outcomes of an experiment.

We can then compare these models to real-life results (e.g. to decide whether a coin is biased).

Definitions

A discrete probability distribution (DPD) only has outcomes that can be listed:
- e.g. Rolling a die gives the outcomes {1,2,3,4,5,6}.
- Each outcome has an assigned probability.

A RV is denoted by a capital letter (usually X).
- A lower-case letter is used to denote a particular outcome of the random variable; the notation P(X = x) means "the probability that the random variable X will have specific outcome x".
- For a fair six-sided die, $P(X = 1) = \frac{1}{6}$.

The expected value (mean) is given by:

$$\mathbb{E}(X) = \sum_x x\, \mathbb{P}(X = x)$$ ← Multiply the value of each outcome by its probability.

A DPD has the following properties:
- The probability of each outcome is between 0 and 1.
- Each outcome is mutually exclusive.
- The probabilities sum to 1.

e.g. The DPD for a biased six-sided die is given as:

$$\mathbb{P}(X = x) = \begin{cases} k \text{ for } x = \{1,2,3,4\} \\ 2k \text{ for } x = \{5,6\} \end{cases}$$

Find the value of k and the expected value.

x	1	2	3	4	5	6
$\mathbb{P}(X = x)$	k	k	k	k	2k	2k

$k + k + k + k + 2k + 2k = 1$ ← Probabilities sum to 1.

$\Rightarrow 8k = 1 \Rightarrow k = \frac{1}{8}$

$\mathbb{E}(X) = 1k + 2k + 3k + 4k + 10k + 12k$

$= 32k = 32 \times \frac{1}{8} = 4$

Experiments with two outcomes

Suppose we have a probability experiment with the following conditions:
- Two possible outcomes - "success" or "fail".
- Each observation or outcome is independent.
- Number of trials, n, is fixed.
- Probability of success, p, is the same for each trial.

For example, our experiment could be rolling a six-sided die a given number of times, n, with the "success" outcome of "rolling a six".

If we let X be the number of successes, X is a discrete RV and is denoted X ~ B(n,p).
- For the die, X is the number of sixes rolled. If we roll the die 10 times, we would write $X \sim B(10, \frac{1}{6})$.

The binomial distribution

If the four criteria above are met, it is appropriate to model X using a binomial distribution:

If $X \sim B(n, p)$ then $\mathbb{P}(X = k) = \binom{n}{k} p^k (1 - p)^{n-k}$

Mean of X = np
Variance of X = np(1 − p)

Probability that X takes specific value k

n choose k (see Binomial Expansion)

e.g. If X is the number of sixes rolled on a fair die and there are 10 rolls (n = 10):

$X \sim B(10, \frac{1}{6})$
The probability of getting two sixes:

$\mathbb{P}(X = 2) = \binom{10}{2} \times (\frac{1}{6})^2 \times (\frac{5}{6})^8$

$= 0.2907$

There are 45 different ways for this combination of successes and failures to occur.

2 "successes" (rolling a six)

8 "fails" (not rolling a six)

4dp is usually sufficient

$X \sim B(10, \frac{1}{6})$
The probability of getting at least one six:
$\mathbb{P}(X \geq 1) = 1 - \mathbb{P}(X = 0)$

$= 1 - \binom{10}{0} \times (\frac{1}{6})^0 \times (\frac{5}{6})^{10}$

$= 1 - 0.1615 = 0.8385$

The mean & variance are calculated as follows:

Mean of X = np = $10 \times \frac{1}{6}$ = 1.67 (3sf)
Variance of X = np(1 − p) = 1.39 (3sf)

Number of expected successes (6s) in 10 rolls

Hypothesis Testing

Applications

Applications

Hypothesis testing is used to decide whether conclusions drawn from a sample can be widely applied to a population.

For example, a supermarket may want to work out whether customers can tell the difference between a named brand and their brand of chocolate. If they ask 20 people and 14 correctly identify the chocolate, can people really tell the difference or did 14 of these customers just "get lucky"?

Key Terms

We study hypothesis testing for the binomial distribution.

▶ Specifically, we make tests on the proportion, p, in $B(n,p)$, where n is known.

▶ In our chocolate example, we assume that X, the number of people who correctly identify the chocolate, has binomial distribution $X \sim B(20, p)$. Can we say with certainty that $p > 0.5$?

The null hypothesis, H_0, is the current situation; the hypothesis that we test.

▶ In the case of the chocolate, this is that people do no better than random chance, i.e. The probability of correctly identifying the chocolate $p = 0.5$.

The alternative hypothesis, H_1, is the alternative claim made

▶ This is the claim that people can tell the difference, so $p > 0.5$ - people do better than "random chance".

The significance level α is the likelihood of incorrectly rejecting H_0, even though it is true.

▶ This is given as a percentage; if we choose to test at 5% significance, this means there is a 5% chance of incorrectly rejecting the null hypothesis.

The test statistic is the assumed distribution with the null hypothesis, $X \sim B(20,0.5)$.

Conducting an hypothesis test

▶ A hypothesis test is conducted in a formulaic way:

1) State null and alternative hypotheses. → H_0: $p = 0.5$
H_1: $p > 0.5$

2) Determine test statistic.

3) Choose significance level.

$X \sim B(20, 0.5)$
Sig. 5% = 0.05

4) Compute value of test statistic:

$$\mathbb{P}(X \geq 14) = 1 - \mathbb{P}(X \leq 13)$$
$$= 1 - 0.9423$$
$$= 0.0577$$

Use ≥ because we are looking for p > 0.5.

Usually given at A Level.

From binomial distribution tables.

5) Make decision and interpret:

$0.0577 > 0.05$
We do not have sufficient evidence to reject H_0. Therefore we conclude that customers cannot tell the difference between the chocolate brands.

Testing at 5% significance level

One- and two-tailed tests

Using H_1: $p > 0.5$ is a one-tailed test - we only test one "tail", or end, of the distribution.

If we instead stated H_1: $p \neq 0.5$ (people could do better or worse than random chance), this would be a two-tailed test, as we would also check $p < 0.5$.

It is important to remember that the critical region is halved; e.g. if we were conducting at 5% significance, would reject H_0 for any test statistic below 0.0025.

Critical values and regions

The critical value is the first value that would cause us to reject H_0.

For $X \sim B(20, 0.5)$ and 5% sig:

x	$\mathbb{P}(X \leq x)$
13	0.9423
14	0.9793
15	0.9941

$P(X \geq 14) = 1 - 0.9423 = 0.0577$
$P(X \geq 15) = 1 - 0.9793 = 0.0207$
X = 15 is a critical value; we reject H_0 for $X \geq 15$.

▶ We also have another critical value at the other end of the distribution:

x	$\mathbb{P}(X \leq x)$
4	0.0059
5	0.0207
6	0.0577

$P(X \leq 4) = 0.0059$
$P(X \leq 5) = 0.0207$
X = 5 is a critical value; we reject H_0 for $X \leq 5$.

The critical region contains all values that would cause us to reject H_0.

▶ Here the regions are $X \leq 5$ and $X \geq 15$.

p-values

▶ The p-value for a hypothesis test is the probability of obtaining a test statistic, X, at least as extreme as th obtained from the sample given that the null hypothe is true. The p-value for our original test is 0.0577.

▶ If a p-value is very close to the significance level α, corresponds to being on the edge of the critical regi as is the case with our test.

▶ The smaller the p-value, the more evidence we have against H_0. If 17 people correctly identify the chocolate, the p-value is 0.0002, giving us very good evidence to reject H_0.